Thermal Engineering

Thermal Engineering

Edited by
Andres Flynn

Larsen & Keller
www.larsen-keller.com

Thermal Engineering
Edited by Andres Flynn
ISBN: 978-1-63549-275-0 (Hardback)

⊟ Larsen & Keller

Published by Larsen and Keller Education,
5 Penn Plaza,
19th Floor,
New York, NY 10001, USA

Cataloging-in-Publication Data

Thermal engineering / edited by Andres Flynn.
    p. cm.
Includes bibliographical references and index.
ISBN 978-1-63549-275-0
1. Heat engineering. 2. Thermodynamics. 3. Heat--Transmission.
I. Flynn, Andres.
TJ265 .T44 2017
621.402 2--dc23

The publisher's policy is to use permanent paper from mills that operate a sustainable forestry policy. Furthermore, the publisher ensures that the text paper and cover boards used have met acceptable environmental accreditation standards.

Printed and bound in the United States of America.

For more information regarding Larsen and Keller Education and its products, please visit the publisher's website www.larsen-keller.com

Table of Contents

Preface

Thermal engineering is an intricate subject, which includes elements from various fields like fluid mechanics, thermodynamics, mass transfer and heat transfer. It is the study of heating and cooling processes, equipment, and theories used in mechanical and chemical engineering. This book presents the complex subject of thermal engineering in the most comprehensible and easy to understand language. Most of the topics introduced in it cover new techniques and the methods of thermal engineering. As this field is emerging at a rapid pace, the contents of this textbook will help the readers understand the modern concepts and applications of the subject.

A short introduction to every chapter is written below to provide an overview of the content of the book:

Chapter 1 - Thermal engineering is the study of processes and equipment that require cooling and heating. The subjects that are considered during a thermal engineering problem are thermodynamics, fluid mechanics and heat transfer. This chapter will provide an integrated understanding of thermal engineering; **Chapter 2** - Thermal energy is the energy present within a system caused due to its temperature. Its not well defined, as the internal energy can be changed without changing the temperature. Thermal fluids, fluid mechanics, combustion and heat transfer are some of the topics covered in this section. This chapter is an overview of the subject matter incorporating all the major aspects of thermal energy; **Chapter 3** - The process that causes the release of energy in any system to its surrounding is known as exothermic process. This release is usually in the form of heat. The exact opposite of exothermic process is endothermic process, while exothermic process releases energy into the environment, endothermic process is the process of absorbing energy from its environment. This section helps the readers in understanding the process related to thermal engineering; **Chapter 4** - In order to develop an in-depth understanding of thermodynamics, it is very important to understand certain topics. Some of these topics are thermodynamics, internal energy, thermodynamic operation, equilibrium thermodynamics and state function. The topics discussed in the chapter are of great importance to broaden the existing knowledge on thermal engineering; **Chapter 5** - Thermodynamics is basically based on four sets of laws. The most prominent law in thermodynamics is the zeroth law of thermodynamics. The additional laws are the first law of thermodynamics, second law of thermodynamics and third law of thermodynamics. This chapter discusses the thermodynamics in a critical manner providing key analysis to the subject matter; **Chapter 6** - Thermal insulation is achievable with the help of engineered methods or processes. It can also be achieved by using suitable objects with apt shapes and materials. The features of thermal insulation that have been covered in the following section are pipe insulation, building insulation and exhaust heat management. The chapter on thermal insulation offers an insightful focus, keeping in mind the complex topic; **Chapter 7 -** The themes discussed in this section are thermodynamic system, thermodynamic process and thermodynamic equilibrium. This section helps the readers in developing an in-depth understanding of the theory of thermal equilibrium. The chapter strategically encompasses and incorporates the key concepts of thermal equilibrium, providing a complete understanding; **Chapter 8 -** Thermal engineering has various features; some of these aspects are thermal conduction, thermal radiation, thermal grease, thermal mass, heat and heat capacity. Thermal conduction is the transmitting of heat by atoms and the movement of electrons within a body. The diverse aspects of thermal engineering in the current scenario have been thoroughly discussed in this chapter.

Finally, I would like to thank my fellow scholars who gave constructive feedback and my family members who supported me at every step.

Editor

Introduction to Thermal Engineering

Thermal engineering is the study of processes and equipment that require cooling and heating. The subjects that are considered during a thermal engineering problem are thermodynamics, fluid mechanics and heat transfer. This chapter will provide an integrated understanding of thermal engineering.

Heating or cooling of processes, equipment, or enclosed environments are within the purview of thermal engineering.

One or more of the following disciplines may be involved in solving a particular thermal engineering problem:

- Thermodynamics

- Fluid mechanics

- Heat transfer

- Mass transfer

Thermal engineering may be practiced by mechanical engineers and chemical engineers.

One branch of knowledge used frequently in thermal engineering is that of thermofluids.

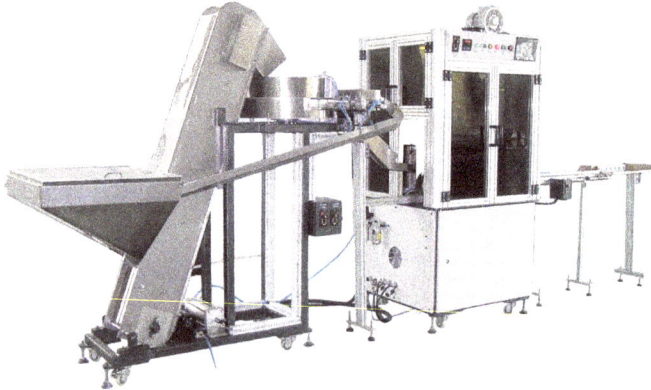

Applications

- Engineering : HVAC

- Cooling of computer chips

- Boiler design

- Solar heating

- In design of combustion engines

- Thermal power plants

- Commercial cooling system

Thermal Energy: An Overview

Thermal energy is the energy present within a system caused due to its temperature. Its not well defined, as the internal energy can be changed without changing the temperature. Thermal fluids, fluid mechanics, combustion and heat transfer are some of the topics covered in this section. This chapter is an overview of the subject matter incorporating all the major aspects of thermal energy.

Thermal Energy

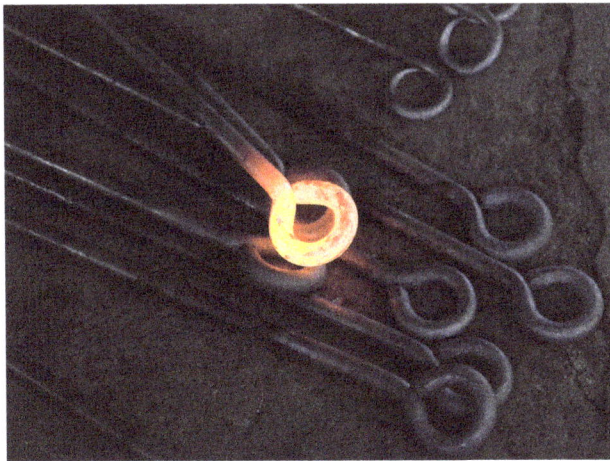

Thermal radiation in visible light can be seen on this hot metalwork. Thermal energy would ideally be the amount of heat required to warm the metal to its temperature, but this quantity is not well-defined, as there are many ways to obtain a given body at a giv-en temperature, and each of them may require a different amount of total heat input. Thermal energy, unlike internal energy, is therefore not a state function.

In thermodynamics, thermal energy refers to the internal energy present in a system due to its temperature. The concept is not well-defined or broadly accepted in physics or thermodynamics, because the internal energy can be changed without changing the temperature, and there is no way to distinguish which part of the internal energy is "thermal". Instead, the appropriate thermodynamic concept is heat, defined as a *transfer* of energy (just as work is another type of transfer of energy). Heat and work therefore depend on the path of transfer and are not state functions, whereas internal energy is a state function.

Relation to Heat and Internal Energy

Heat is energy transferred spontaneously from a hotter to a colder system or body. Heat is energy in transfer, not a property of the system; it is not *contained* within the boundary of the system. On the other hand, internal energy is a property of a system, and exists on both sides of a boundary. In an ideal gas, the internal energy is the statistical mean of the kinetic energy of the gas particles, and it is this kinetic motion that is the source and the effect of the transfer of heat across a system boundary. The internal energy of an ideal gas can in this sense be regarded as "thermal energy". In this case, however, thermal energy and internal energy are identical.

Systems that are more complex than ideal gases (such as real gases) can undergo phase transitions. Phase transitions can change the internal energy of the system without changing its temperature. Therefore, the thermal energy cannot be defined solely by the temperature. Thermal energy also cannot be defined by the difference between the internal energy and the net heat transfer in or out of the system, because it is easy to construct thermodynamic cycles where the system begins and ends in exactly the same state, but there is a net flow of heat in or out during the cycle.

For these reasons, the concept of the thermal energy of a system is ill-defined and is not used in thermodynamics.

Historical Context

In an 1847 lecture entitled *On Matter, Living Force, and Heat,* James Prescott Joule characterized various terms that are closely related to thermal energy and heat. He identified the terms latent heat and sensible heat as forms of heat each effecting distinct physical phenomena, namely the potential and kinetic energy of particles, respectively. He describes latent energy as the energy of interaction in a given configuration of particles, i.e. a form of potential energy, and the sensible heat as an energy affecting temperature measured by the thermometer due to the thermal energy, which he called the *living force*.

Thermal Fluids

Thermal fluids or Thermofluids is a branch of science and engineering divided into four sections:

- Heat transfer

- Thermodynamics

- Fluid mechanics

- Combustion

The term consists of two words: "thermal", meaning heat, and "fluids", which refers to liquids, gases and vapors. Pressure, volume, and density all play an important role in thermal fluids. Temperature, flow rate, phase transition and chemical reactions may also be important in a thermal fluids context.

Heat Transfer

Heat transfer is a discipline of thermal engineering that concerns the transfer of thermal energy from one physical system to another. Heat transfer is classified into various mechanisms, such as heat conduction, convection, thermal radiation, and phase-change transfer. Engineers also consider the transfer of mass of differing chemical species, either cold or hot, to achieve heat transfer.

Sections include :

- Energy transfer by heat, work and mass
- Laws of thermodynamics
- Entropy
- Refrigeration Techniques
- Properties and nature of pure substances

Applications

- Engineering : Predicting and analysing the performance of machines

Thermodynamics

Thermodynamics is the science of energy conversion involving heat and other forms of energy, most notably mechanical work. It studies and interrelates the macroscopic variables, such as temperature, volume and pressure, which describe physical, thermodynamic systems.

Fluid Mechanics

Fluid Mechanics the study of the physical forces at work during fluid flow. Fluid mechanics can be divided into fluid kinematics, the study of fluid motion, and fluid dynamics, the study of the effect of forces on fluid motion, which can further be divided into fluid statics, the study of fluids at rest, and fluid kinetics, the study of fluids in motion. Some of its more interesting concepts include momentum and reactive forces in fluid flow and fluid machinery theory and performance.

Sections include:

- Fluid flow and continuity

- Momentum in fluids

- Static and dynamic forces on a boundary

- Laminar and turbulent flow

- Metacentric height and vessel stability

Applications

- Pump Design.

- Hydro-Electric Power Generation.

- Naval Architecture.

Combustion

Combustion is the sequence of exothermic chemical reactions between a fuel and an oxidant accompanied by the production of heat and conversion of chemical species. The release of heat can result in the production of light in the form of either glowing or a flame. Fuels of interest often include organic compounds (especially hydrocarbons) in the gas, liquid or solid phase.

Fluid Mechanics

Fluid mechanics is the branch of physics that studies the mechanics of fluids (liquids, gases, and plasmas) and the forces on them. Fluid mechanics has a wide range of applications, including for mechanical engineering, chemical engineering, geophysics, astrophysics, and biology. Fluid mechanics can be divided into fluid statics, the study of fluids at rest; and fluid dynamics, the study of the effect of forces on fluid motion. It is a branch of continuum mechanics, a subject which models matter without using the information that it is made out of atoms; that is, it models matter from a *macroscopic* viewpoint rather than from *microscopic*. Fluid mechanics, especially fluid dynamics, is an active field of research with many problems that are partly or wholly unsolved. Fluid mechanics can be mathematically complex, and can best be solved by numerical methods, typically using computers. A modern discipline, called computational fluid dynamics (CFD), is devoted to this approach to solving fluid mechanics problems. Particle image velocimetry, an experimental method for visualizing and analyzing fluid flow, also takes advantage of the highly visual nature of fluid flow.

Brief History

The study of fluid mechanics goes back at least to the days of ancient Greece, when

Archimedes investigated fluid statics and buoyancy and formulated his famous law known now as the Archimedes' principle, which was published in his work *On Floating Bodies* – generally considered to be the first major work on fluid mechanics. Rapid advancement in fluid mechanics began with Leonardo da Vinci (observations and experiments), Evangelista Torricelli (invented the barometer), Isaac Newton (investigated viscosity) and Blaise Pascal (researched hydrostatics, formulated Pascal's law), and was continued by Daniel Bernoulli with the introduction of mathematical fluid dynamics in *Hydrodynamica* (1738).

Inviscid flow was further analyzed by various mathematicians (Leonhard Euler, Jean le Rond d'Alembert, Joseph Louis Lagrange, Pierre-Simon Laplace, Siméon Denis Poisson) and viscous flow was explored by a multitude of engineers including Jean Léonard Marie Poiseuille and Gotthilf Hagen. Further mathematical justification was provided by Claude-Louis Navier and George Gabriel Stokes in the Navier–Stokes equations, and boundary layers were investigated (Ludwig Prandtl, Theodore von Kármán), while various scientists such as Osborne Reynolds, Andrey Kolmogorov, and Geoffrey Ingram Taylor advanced the understanding of fluid viscosity and turbulence.

Main Branches

Fluid Statics

Fluid statics or hydrostatics is the branch of fluid mechanics that studies fluids at rest. It embraces the study of the conditions under which fluids are at rest in stable equilibrium; and is contrasted with fluid dynamics, the study of fluids in motion. Hydrostatics offers physical explanations for many phenomena of everyday life, such as why atmospheric pressure changes with altitude, why wood and oil float on water, and why the surface of water is always flat and horizontal whatever the shape of its container. Hydrostatics is fundamental to hydraulics, the engineering of equipment for storing, transporting and using fluids. It is also relevant to some aspect of geophysics and astrophysics (for example, in understanding plate tectonics and anomalies in the Earth's gravitational field), to meteorology, to medicine (in the context of blood pressure), and many other fields.

Fluid Dynamics

Fluid dynamics is a subdiscipline of fluid mechanics that deals with fluid flow—the science of liquids and gases in motion. Fluid dynamics offers a systematic structure—which underlies these practical disciplines—that embraces empirical and semi-empirical laws derived from flow measurement and used to solve practical problems. The solution to a fluid dynamics problem typically involves calculating various properties of the fluid, such as velocity, pressure, density, and temperature, as functions of space and time. It has several subdisciplines itself, including aerodynamics (the study of air and other gases in motion) and hydrodynamics (the study of liquids in motion). Fluid dynamics has a wide range of applications, including calculating forces and moments on aircraft, determining the mass flow rate of petroleum through pipelines, predicting evolving weather patterns, understanding nebulae in interstellar space and modeling explosions. Some fluid-dynamical principles are used in traffic engineering and crowd dynamics.

Relationship to Continuum Mechanics

Fluid mechanics is a subdiscipline of continuum mechanics, as illustrated in the following table.

Continuum mechanics The study of the physics of continuous materials	Solid mechanics The study of the physics of continuous materials with a defined rest shape.	Elasticity Describes materials that return to their rest shape after applied stresses are removed.	
		Plasticity Describes materials that permanently deform after a sufficient applied stress.	Rheology The study of materials with both solid and fluid characteristics.
	Fluid mechanics The study of the physics of continuous materials which deform when subjected to a force.	Non-Newtonian fluids do not undergo strain rates proportional to the applied shear stress.	
		Newtonian fluids undergo strain rates proportional to the applied shear stress.	

In a mechanical view, a fluid is a substance that does not support shear stress; that is why a fluid at rest has the shape of its containing vessel. A fluid at rest has no shear stress.

Assumptions

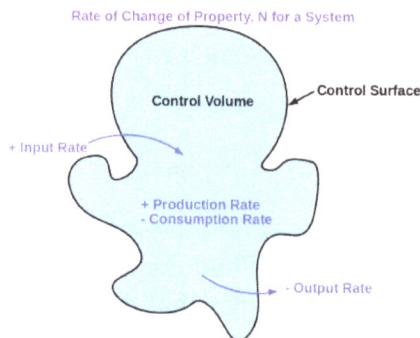

Balance for some integrated fluid quantity in a control volume enclosed by a control surface.

The assumptions inherent to a fluid mechanical treatment of a physical system can be expressed in terms of mathematical equations. Fundamentally, every fluid mechanical system is assumed to obey:

- Conservation of mass

- Conservation of energy

- Conservation of momentum

- The *continuum assumption*

For example, the assumption that mass is conserved means that for any fixed control volume (for example, a spherical volume) – enclosed by a control surface – the rate of change of the mass contained in that volume is equal to the rate at which mass is passing through the surface from *outside* to *inside*, minus the rate at which mass is passing from *inside* to *outside*. This can be expressed as an equation in integral form over the control volume.

The continuum assumption is an idealization of continuum mechanics under which fluids can be treated as continuous, even though, on a microscopic scale, they are composed of molecules. Under the continuum assumption, macroscopic (observed/measurable) properties such as density, pressure, temperature, and bulk velocity are taken to be well-defined at "infinitesimal" volume elements -- small in comparison to the characteristic length scale of the system, but large in comparison to molecular length scale. Fluid properties can vary continuously from one volume element to another and are average values of the molecular properties. The continuum hypothesis can lead to inaccurate results in applications like supersonic speed flows, or molecular flows on nano scale. Those problems for which the continuum hypothesis fails, can be solved using statistical mechanics. To determine whether or not the continuum hypothesis applies, the Knudsen number, defined as the ratio of the molecular mean free path to the characteristic length scale, is evaluated. Problems with Knudsen numbers below 0.1 can be evaluated using continuum hypothesis, but molecular approach (statistical mechanics) can be applied for all ranges of Knudsen numbers.

Navier–Stokes Equations

The Navier–Stokes equations (named after Claude-Louis Navier and George Gabriel Stokes) are differential equations that describe the force balance at a given point within a fluid. For an incompressible fluid with vector velocity field $\mathbf{u}$, the Navier–Stokes equations are

$$\frac{\partial \mathbf{u}}{\partial t} + (\mathbf{u} \cdot \nabla)\mathbf{u} = -\frac{1}{\rho}\nabla P + \nu \nabla^2 \mathbf{u},$$

where $\mathbf{u}$ is the flow velocity vector. Analogous to Newton's equations of motion, the Navier–Stokes equations describe changes in momentum (force) in response to pressure P and viscosity, parameterized, here, by the kinematic viscosity ν. Occasionally, body forces, such as the gravitational force or Lorentz force are added to the equations.

Solutions of the Navier–Stokes equations for a given physical problem must be sought with the help of calculus. In practical terms only the simplest case can be solved exactly in this way. These cases generally involve non-turbulent, steady flow in which the Reynolds number is small. For more complex cases, especially those involving turbulence, such as global weather systems, aerodynamics, hydrodynamics and many more, solutions of the Navier–Stokes equations can currently only be found with the help of computers. This branch of science is called computational fluid dynamics.

Inviscid and Viscous Fluids

An inviscid fluid has no viscosity, $v = 0$. In practice, an inviscid flow is an idealization, one that facilitates mathematical treatment. In fact, purely inviscid flows are only known to be realized in the case of superfluidity. Otherwise, fluids are generally **viscous**, a property that is often most important within a boundary layer near a solid surface, where the flow must match onto the no-slip condition at the solid. In some cases, the mathematics of a fluid mechanical system can be treated by assuming that the fluid outside of boundary layers is inviscid, and then matching its solution onto that for a thin laminar boundary layer.

For fluid flow over a porous boundary, the fluid velocity can be discontinuous between the free fluid and the fluid in the porous media (this is related to the Beavers and Joseph condition). Further, it is useful at low subsonic speeds to assume that a gas is incompressible — that is, the density of the gas does not change even though the speed and static pressure change.

Newtonian Versus Non-Newtonian Fluids

A Newtonian fluid (named after Isaac Newton) is defined to be a fluid whose shear stress is linearly proportional to the velocity gradient in the direction perpendicular to the plane of shear. This definition means regardless of the forces acting on a fluid, it *continues to flow*. For example, water is a Newtonian fluid, because it continues to display fluid properties no matter how much it is stirred or mixed. A slightly less rigorous definition is that the drag of a small object being moved slowly through the fluid is proportional to the force applied to the object. (Compare friction). Important fluids, like water as well as most gases, behave – to good approximation – as a Newtonian fluid under normal conditions on Earth.

By contrast, stirring a non-Newtonian fluid can leave a "hole" behind. This will gradually fill up over time – this behaviour is seen in materials such as pudding, oobleck, or sand (although sand isn't strictly a fluid). Alternatively, stirring a non-Newtonian fluid can cause the viscosity to decrease, so the fluid appears "thinner" (this is seen in non-drip paints). There are many types of non-Newtonian fluids, as they are defined to be something that fails to obey a particular property – for example, most fluids with long molecular chains can react in a non-Newtonian manner.

Equations for a Newtonian Fluid

The constant of proportionality between the viscous stress tensor and the velocity gradient is known as the viscosity. A simple equation to describe incompressible Newtonian fluid behaviour is

$$\tau = -\mu \frac{dv}{dy}$$

where

> τ is the shear stress exerted by the fluid ("drag")
>
> μ is the fluid viscosity – a constant of proportionality
>
> $\frac{dv}{dy}$ Sis the velocity gradient perpendicular to the direction of shear.

For a Newtonian fluid, the viscosity, by definition, depends only on temperature and pressure, not on the forces acting upon it. If the fluid is incompressible the equation governing the viscous stress (in Cartesian coordinates) is

$$\tau_{ij} = \mu \left(\frac{\partial v_i}{\partial x_j} + \frac{\partial v_j}{\partial x_i} \right)$$

where

> τ_{ij} is the shear stress on the ith face of a fluid element in the jth direction
>
> v_i is the velocity in the ith direction
>
> x_j is the jth direction coordinate.

If the fluid is not incompressible the general form for the viscous stress in a Newtonian fluid is

$$\tau_{ij} = \mu \left(\frac{\partial v_i}{\partial x_j} + \frac{\partial v_j}{\partial x_i} - \frac{2}{3} \delta_{ij} \nabla \cdot \mathbf{v} \right) + \kappa \delta_{ij} \nabla \cdot \mathbf{v}$$

where κ is the second viscosity coefficient (or bulk viscosity). If a fluid does not obey this relation, it is termed a non-Newtonian fluid, of which there are several types. Non-Newtonian fluids can be either plastic, Bingham plastic, pseudoplastic, dilatant, thixotropic, rheopectic, viscoelastic.

In some applications another rough broad division among fluids is made: ideal and non-ideal fluids. An Ideal fluid is non-viscous and offers no resistance whatsoever to a shearing force. An ideal fluid really does not exist, but in some calculations, the assumption is justifiable. One example of this is the flow far from solid surfaces. In many cases the viscous effects are concentrated near the solid boundaries (such as in boundary layers) while in regions of the flow field far away from the boundaries the viscous effects can be neglected and the fluid there is treated as it were inviscid (ideal flow).

Combustion

The flames caused as a result of a fuel undergoing combustion (burning)

Combustion or burning is a high-temperature exothermic redox chemical reaction between a fuel and an oxidant, usually atmospheric oxygen, that produces oxidized, often gaseous products, in a mixture termed as smoke. Combustion in a fire produces a flame, and the heat produced can make combustion self-sustaining. Combustion is often a complicated sequence of elementary radical reactions. Solid fuels, such as wood, first undergo endothermic pyrolysis to produce gaseous fuels whose combustion then supplies the heat required to produce more of them. Combustion is often hot enough that light in the form of either glowing or a flame is produced. A simple example can be seen in the combustion of hydrogen and oxygen into water vapor, a reaction commonly used to fuel rocket engines. This reaction releases 242 kJ/mol of heat and reduces the enthalpy accordingly (at constant temperature and pressure):

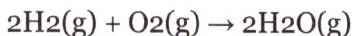

$$2H2(g) + O2(g) \rightarrow 2H2O(g)$$

Combustion of an organic fuel in air is always exothermic because the double bond in O_2 is much weaker than other double bonds or pairs of single bonds, and therefore the formation of the stronger bonds in the combustion products CO_2 and H_2O results in the release of energy. The bond energies in the fuel play only a minor role, since they are similar to those in the combustion products; e.g., the sum of the bond energies of CH_4 is nearly the same as that of CO_2. The heat of combustion is approximately -418 kJ per mole of O_2 used up in the combustion reaction, and can be estimated from the elemental composition of the fuel.

Uncatalyzed combustion in air requires fairly high temperatures. Complete combustion is stoichiometric with respect to the fuel, where there is no remaining fuel, and ideally, no remaining oxidant. Thermodynamically, the chemical equilibrium of combustion in air is overwhelmingly on the side of the products. However, complete combustion is almost impossible to achieve, since the chemical equilibrium is not necessarily reached, or may contain unburnt products such as carbon monoxide, hydrogen and even carbon (soot or ash). Thus, the produced smoke is usually toxic and contains unburned or partially oxidized products. Any combustion at high temperatures in atmospheric air, which is 78 percent nitrogen, will also create small amounts of several nitrogen oxides, commonly referred to as NO_x, since the combustion of nitrogen is thermodynamically favored at high, but not low temperatures. Since combustion is rarely clean, flue gas cleaning or catalytic converters may be required by law.

Fires occur naturally, ignited by lightning strikes or by volcanic products. Combustion (fire) was the first controlled chemical reaction discovered by humans, in the form of campfires and bonfires, and continues to be the main method to produce energy for humanity. Usually, the fuel is carbon, hydrocarbons or more complicated mixtures such as wood that contains partially oxidized hydrocarbons. The thermal energy produced from combustion of either fossil fuels such as coal or oil, or from renewable fuels such as firewood, is harvested for diverse uses such as cooking, production of electricity or industrial or domestic heating. Combustion is also currently the only reaction used to power rockets. Combustion is also used to destroy (incinerate) waste, both nonhazardous and hazardous.

Oxidants for combustion have high oxidation potential and include atmospheric or pure oxygen, chlorine, fluorine, chlorine trifluoride, nitrous oxide and nitric acid. For instance, hydrogen burns in chlorine to form hydrogen chloride with the liberation of heat and light characteristic of combustion. Although usually not catalyzed, combustion can be catalyzed by platinum or vanadium, as in the contact process.

Types

Complete vs. Incomplete

Complete

In complete combustion, the reactant burns in oxygen, producing a limited number of products. When a hydrocarbon burns in oxygen, the reaction will primarily yield carbon dioxide and water. When elements are burned, the products are primarily the most common oxides. Carbon will yield carbon dioxide, sulfur will yield sulfur dioxide, and iron will yield iron(III) oxide. Nitrogen is not considered to be a combustible substance when oxygen is the oxidant, but small amounts of various nitrogen oxides (commonly designated NO x species) form when air is the oxidant.

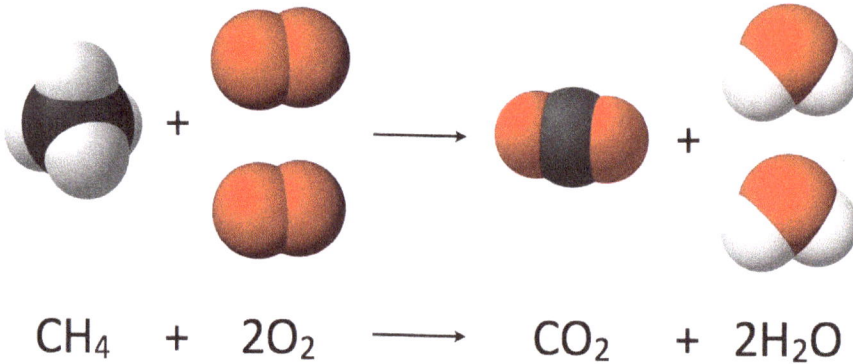

The combustion of methane, a hydrocarbon.

Combustion is not necessarily favorable to the maximum degree of oxidation, and it can be temperature-dependent. For example, sulfur trioxide is not produced quantitatively by the combustion of sulfur. NO_x species appear in significant amounts above about 2,800 °F (1,540 °C), and more is produced at higher temperatures. The amount of NO_x is also a function of oxygen excess.

In most industrial applications and in fires, air is the source of oxygen (O_2). In air, each mole of oxygen is mixed with approximately 3.71 mol of nitrogen. Nitrogen does not take part in combustion, but at high temperatures some nitrogen will be converted to NO_x (mostly NO, with much smaller amounts of NO_2). On the other hand, when there is insufficient oxygen to completely combust the fuel, some fuel carbon is converted to carbon monoxide and some of the hydrogen remains unreacted. A more complete set of equations for the combustion of a hydrocarbon in air therefore requires an additional calculation for the distribution of oxygen between the carbon and hydrogen in the fuel.

The amount of air required for complete combustion to take place is known as theoretical air. However, in practice the air used is 2-3x that of theoretical air.

Incomplete

Incomplete combustion will occur when there is not enough oxygen to allow the fuel to react completely to produce carbon dioxide and water. It also happens when the combustion is quenched by a heat sink, such as a solid surface or flame trap.

For most fuels, such as diesel oil, coal or wood, pyrolysis occurs before combustion. In incomplete combustion, products of pyrolysis remain unburnt and contaminate the smoke with noxious particulate matter and gases. Partially oxidized compounds are also a concern; partial oxidation of ethanol can produce harmful acetaldehyde, and carbon can produce toxic carbon monoxide.

The quality of combustion can be improved by the designs of combustion devices, such as burners and internal combustion engines. Further improvements are achievable by catalytic after-burning devices (such as catalytic converters) or by the simple partial return of the exhaust gases into the combustion process. Such devices are required by environmental legislation for cars in most countries, and may be necessary to enable large combustion devices, such as thermal power stations, to reach legal emission standards.

The degree of combustion can be measured and analyzed with test equipment. HVAC contractors, firemen and engineers use combustion analyzers to test the efficiency of a burner during the combustion process. In addition, the efficiency of an internal combustion engine can be measured in this way, and some U.S. states and local municipalities use combustion analysis to define and rate the efficiency of vehicles on the road today.

Smouldering

Smouldering is the slow, low-temperature, flameless form of combustion, sustained by the heat evolved when oxygen directly attacks the surface of a condensed-phase fuel. It is a typically incomplete combustion reaction. Solid materials that can sustain a smouldering reaction include coal, cellulose, wood, cotton, tobacco, peat, duff, humus, synthetic foams, charring polymers (including polyurethane foam) and dust. Common examples of smouldering phenomena are the initiation of residential fires on upholstered furniture by weak heat sources (e.g., a cigarette, a short-circuited wire) and the persistent combustion of biomass behind the flaming fronts of wildfires.

Rapid

An experiment that demonstrates the large amount of energy released upon combustion of ethanol. A mixture of alcohol (in this case, ethanol) vapour and air in a large plastic bottle with a small neck is ignited, resulting in a large blue flame and a 'whoosh' sound.

Rapid combustion is a form of combustion, otherwise known as a fire, in which large amounts of heat and light energy are released, which often results in a flame. This is used in a form of machinery such as internal combustion engines and in thermobaric weapons. Such a combustion is frequently called an explosion, though for an internal combustion engine this is inaccurate. An internal combustion engine nominally operates on a controlled rapid burn. When the fuel-air mixture in an internal combustion engine explodes, that is known as detonation.

Spontaneous

Spontaneous combustion is a type of combustion which occurs by self heating (increase in temperature due to exothermic internal reactions), followed by thermal runaway (self heating which rapidly accelerates to high temperatures) and finally, ignition. For example, phosphorus self-ignites at room temperature without the application of heat.

Turbulent

Combustion resulting in a turbulent flame is the most used for industrial application (e.g. gas turbines, gasoline engines, etc.) because the turbulence helps the mixing process between the fuel and oxidizer.

Micro-gravity

Colourized gray-scale composite image of the individual frames from a video of a backlit fuel droplet burning in microgravity.

The term 'micro' gravity refers to a gravitational state that is 'low' (i.e., 'micro' in the sense of 'small' and not necessarily a millionth of Earth's normal gravity) such that the influence of buoyancy on physical processes may be considered small relative to other flow processes that would be present at normal gravity. In such an environment, the thermal and flow transport dynamics can behave quite differently than in normal gravity conditions (e.g., a candle's flame takes the shape of a sphere.). Microgravity combustion research contributes to the understanding of a wide variety of aspects that are relevant to both the environment of a spacecraft (e.g., fire dynamics relevant to crew safety on the International Space Station) and terrestrial (Earth-based) conditions (e.g., droplet combustion dynamics to assist developing new fuel blends for improved combustion, materials fabrication processes, thermal management of electronic systems, multiphase flow boiling dynamics, and many others).

Micro-combustion

Combustion processes which happen in very small volumes are considered micro-combustion. The high surface-to-volume ratio increases specific heat loss. Quenching distance plays a vital role in stabilizing the flame in such combustion chambers.

Fuels

Substances or materials which undergo combustion are called fuels. The most common examples are natural gas, propane, kerosene, diesel, petrol, charcoal, coal, wood, etc.

Liquid Fuels

Combustion of a liquid fuel in an oxidizing atmosphere actually happens in the gas phase. It is the vapor that burns, not the liquid. Therefore, a liquid will normally catch fire only above a certain temperature: its flash point. The flash point of a liquid fuel is the lowest temperature at which it can form an ignitable mix with air. It is the minimum temperature at which there is enough evaporated fuel in the air to start combustion.

Solid Fuels

The act of combustion consists of three relatively distinct but overlapping phases:

- Preheating phase, when the unburned fuel is heated up to its flash point and then fire point. Flammable gases start being evolved in a process similar to dry distillation.

- Distillation phase or gaseous phase, when the mix of evolved flammable gases with oxygen is ignited. Energy is produced in the form of heat and light. Flames are often visible. Heat transfer from the combustion to the solid maintains the evolution of flammable vapours.

- Charcoal phase or solid phase, when the output of flammable gases from the material is too low for persistent presence of flame and the charred fuel does not burn rapidly and just glows and later only smoulders.

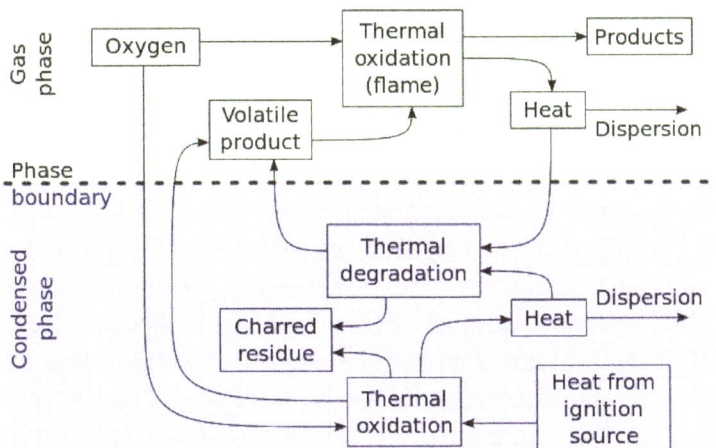

A general scheme of polymer combustion

Combustion Management

Efficient process heating requires recovery of the largest possible part of a fuel's heat of combustion into the material being processed. There are many avenues of loss in the operation of a heating process. Typically, the dominant loss is sensible heat leaving with the offgas (i.e., the flue gas). The temperature and quantity of offgas indicates its heat content (enthalpy), so keeping its quantity low minimizes heat loss.

In a perfect furnace, the combustion air flow would be matched to the fuel flow to give each fuel molecule the exact amount of oxygen needed to cause complete combustion. However, in the real world, combustion does not proceed in a perfect manner. Unburned fuel (usually CO and H2) discharged from the system represents a heating value loss (as well as a safety hazard). Since combustibles are undesirable in the offgas, while the presence of unreacted oxygen there presents minimal safety and environmental concerns, the first principle of combustion management is to provide more oxygen than is theoretically needed to ensure that all the fuel burns. For methane (CH_4) combustion, for example, slightly more than two molecules of oxygen are required.

The second principle of combustion management, however, is to not use too much oxygen. The correct amount of oxygen requires three types of measurement: first, active control of air and fuel flow; second, offgas oxygen measurement; and third, measurement of offgas combustibles. For each heating process there exists an optimum condition of minimal offgas heat loss with acceptable levels of combustibles concentration. Minimizing excess oxygen pays an additional benefit: for a given offgas temperature, the NOx level is lowest when excess oxygen is kept lowest.

Adherence to these two principles is furthered by making material and heat balances on the combustion process. The material balance directly relates the air/fuel ratio to the percentage of O2 in the combustion gas. The heat balance relates the heat available for the charge to the overall net heat produced by fuel combustion. Additional material and heat balances can be made to quantify the thermal advantage from preheating the combustion air, or enriching it in oxygen.

Reaction Mechanism

Combustion in oxygen is a chain reaction in which many distinct radical intermediates participate. The high energy required for initiation is explained by the unusual structure of the dioxygen molecule. The lowest-energy configuration of the dioxygen molecule is a stable, relatively unreactive diradical in a triplet spin state. Bonding can be described with three bonding electron pairs and two antibonding electrons, whose spins are aligned, such that the molecule has nonzero total angular momentum. Most fuels, on the other hand, are in a singlet state, with paired spins and zero total angular momentum. Interaction between the two is quantum mechanically a "forbidden transition", i.e. possible with a very low probability. To initiate combustion, energy is required to force dioxygen into a spin-paired state, or singlet oxygen. This intermediate is extremely reactive. The energy is supplied as heat, and the reaction then produces additional heat, which allows it to continue.

Combustion of hydrocarbons is thought to be initiated by hydrogen atom abstraction (not proton abstraction) from the fuel to oxygen, to give a hydroperoxide radical (HOO). This reacts further to give hydroperoxides, which break up to give hydroxyl radicals. There are a great variety of these processes that produce fuel radicals and oxidizing radicals. Oxidizing species include singlet oxygen, hydroxyl, monatomic oxygen, and hydroperoxyl. Such intermediates are short-lived and cannot be isolated. However, non-radical intermediates are stable and are produced in incomplete combustion. An example is acetaldehyde produced in the combustion of ethanol. An intermediate in the combustion of carbon and hydrocarbons, carbon monoxide, is of special importance because it is a poisonous gas, but also economically useful for the production of syngas.

Solid and heavy liquid fuels also undergo a great number of pyrolysis reactions that give more easily oxidized, gaseous fuels. These reactions are endothermic and require constant energy input from the ongoing combustion reactions. A lack of oxygen or other poorly designed conditions result in these noxious and carcinogenic pyrolysis products being emitted as thick, black smoke.

The rate of combustion is the amount of a material that undergoes combustion over a period of time. It can be expressed in grams per second (g/s) or kilograms per second (kg/s).

Detailed descriptions of combustion processes, from the chemical kinetics perspective, requires the formulation of large and intricate webs of elementary reactions. For instance, combustion of hydrocarbon fuels typically involve hundreds of chemical species reacting according to thousands of reactions.

Inclusion of such mechanisms within computational flow solvers still represents a pretty challenging task mainly in two aspects. First, the number of degrees of freedom (proportional to the number of chemical species) can be dramatically large; second the source term due to reactions introduces a disparate number of time scales which makes the whole dynamical system stiff. As a result, the direct numerical simulation of turbulent reactive flows with heavy fuels soon becomes intractable even for modern supercomputers.

Therefore, a plethora of methodologies has been devised for reducing the complexity of combustion mechanisms without renouncing to high detail level. Examples are provided by: the Relaxation Redistribution Method (RRM) The Intrinsic Low-Dimensional Manifold (ILDM) approach and further developments The invariant constrained equilibrium edge preimage curve method. A few variational approaches The Computational

Singular perturbation (CSP) method and further developments. The Rate Controlled Constrained Equilibrium (RCCE) and Quasi Equilibrium Manifold (QEM) approach. The G-Scheme. The Method of Invariant Grids (MIG).

Temperature

Antoine Lavoisier conducting an experiment related combustion generated by amplified sun light.

Assuming perfect combustion conditions, such as complete combustion under adiabatic conditions (i.e., no heat loss or gain), the adiabatic combustion temperature can be determined. The formula that yields this temperature is based on the first law of thermodynamics and takes note of the fact that the heat of combustion is used entirely for heating the fuel, the combustion air or oxygen, and the combustion product gases (commonly referred to as the *flue gas*).

In the case of fossil fuels burnt in air, the combustion temperature depends on all of the following:

The adiabatic combustion temperature (also known as the *adiabatic flame temperature*) increases for higher heating values and inlet air and fuel temperatures and for stoichiometric air ratios approaching one.

Most commonly, the adiabatic combustion temperatures for coals are around 2,200 °C (3,992 °F) (for inlet air and fuel at ambient temperatures), around 2,150 °C (3,902 °F) for oil and 2,000 °C (3,632 °F) for natural gas.

In industrial fired heaters, power stationsteam generators, and large gas-fired turbines, the more common way of expressing the usage of more than the stoichiometric combustion air is *percent excess combustion air*. For example, excess combustion air of 15 percent means that 15 percent more than the required stoichiometric air is being used.

Instabilities

Combustion instabilities are typically violent pressure oscillations in a combustion chamber. These pressure oscillations can be as high as 180 dB, and long term exposure to these cyclic pressure and thermal loads reduces the life of engine components. In rockets, such as the F1 used in the Saturn V program, instabilities led to massive damage of the combustion chamber and surrounding components. This problem was solved by re-designing the fuel injector. In liquid jet engines the droplet size and distribution can be used to attenuate the instabilities. Combustion instabilities are a major concern in ground-based gas turbine engines because of NOx emissions. The tendency is to run lean, an equivalence ratio less than 1, to reduce the combustion temperature and thus reduce the NOx emissions; however, running the combustion lean makes it very susceptible to combustion instability.

Heat Transfer

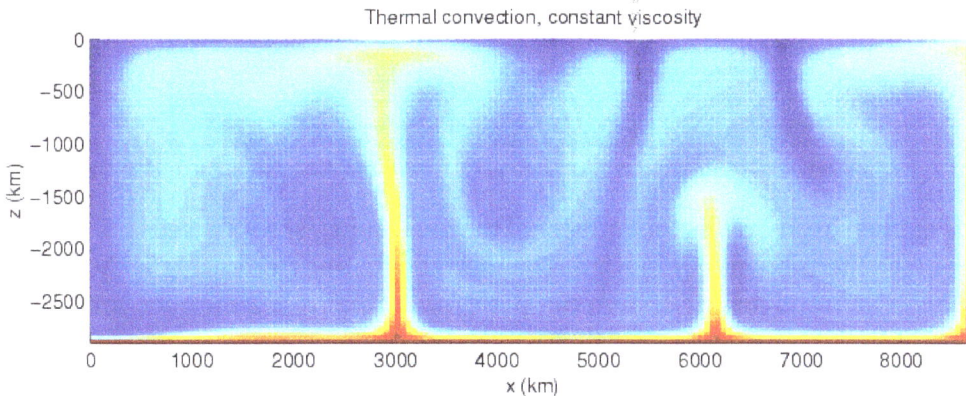

This figure shows a calculation for thermal convection in the Earth's mantle. Colors closer to red are hot areas and colors closer to blue are cold areas. A hot, less-dense lower boundary layer sends plumes of hot material upwards, and likewise, cold material from the top moves downwards.

Heat transfer is the exchange of thermal energy between physical systems. The rate of heat transfer is dependent on the temperatures of the systems and the properties of the intervening medium through which the heat is transferred. The three fundamental modes of heat transfer are *conduction*, *convection* and *radiation*. Heat transfer, the flow of energy in the form of heat, is a process by which a system's internal energy is changed, hence is of vital use in applications of the First Law of Thermodynamics. Conduction is also known as diffusion.

The direction of heat transfer is from a region of high temperature to another region of lower temperature, and is governed by the Second Law of Thermodynamics. Heat transfer changes the internal energy of the systems from which and to which the energy is transferred. Heat transfer will occur in a direction that increases the entropy of the collection of systems.

Heat transfer ceases when thermal equilibrium is reached, at which point all involved bodies and the surroundings reach the same temperature. Thermal expansion is the tendency of matter to change in volume in response to a change in temperature.

Overview

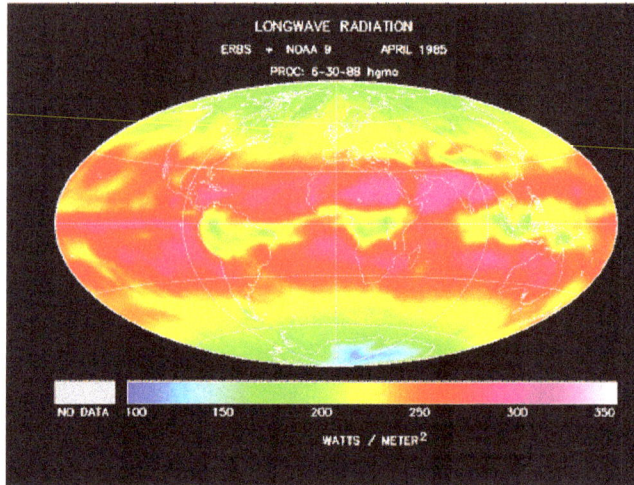

Earth's longwave thermal radiation intensity, from clouds, atmosphere and surface.

Heat is defined in physics as the transfer of thermal energy across a well-defined boundary around a thermodynamic system. The thermodynamic free energy is the amount of work that a thermodynamic system can perform. Enthalpy is a thermodynamic potential, designated by the letter "H", that is the sum of the internal energy of the system (U) plus the product of pressure (P) and volume (V). Joule is a unit to quantify energy, work, or the amount of heat.

Heat transfer is a. process function (or path function), as opposed to functions of state; therefore, the amount of heat transferred in a thermodynamic process that changes the state of a system depends on how that process occurs, not only the net difference between the initial and final states of the process.

Thermodynamic and mechanical heat transfer is calculated with the heat transfer coefficient, the proportionality between the heat flux and the thermodynamic driving force for the flow of heat. Heat flux is a quantitative, vectorial representation of heat-flow through a surface.

In engineering contexts, the term *heat* is taken as synonymous to thermal energy. This usage has its origin in the historical interpretation of heat as a fluid (*caloric*) that can be transferred by various causes, and that is also common in the language of laymen and everyday life.

The transport equations for thermal energy (Fourier's law), mechanical momentum (Newton's law for fluids), and mass transfer (Fick's laws of diffusion) are similar, and analogies among these three transport processes have been developed to facilitate prediction of conversion from any one to the others.

Thermal engineering concerns the generation, use, conversion, and exchange of heat transfer. As such, heat transfer is involved in almost every sector of the economy. Heat transfer is classified into various mechanisms, such as thermal conduction, thermal convection, thermal radiation, and transfer of energy by phase changes.

Transient conduction occurs when the temperature within an object changes as a function of time. Analysis of transient systems is more complex and often calls for the application of approximation theories or numerical analysis by computer.

Convection

The flow of fluid may be forced by external processes, or sometimes (in gravitational fields) by buoyancy forces caused when thermal energy expands the fluid (for example in a fire plume), thus influencing its own transfer. The latter process is often called "natural convection". All convective processes also move heat partly by diffusion, as well. Another form of convection is forced convection. In this case the fluid is forced to flow by use of a pump, fan or other mechanical means.

Convective heat transfer, or convection, is the transfer of heat from one place to another by the movement of fluids, a process that is essentially the transfer of heat via mass transfer. Bulk motion of fluid enhances heat transfer in many physical situations, such as (for example) between a solid surface and the fluid. Convection is usually the dominant form of heat transfer in liquids and gases. Although sometimes discussed as a third method of heat transfer, convection is usually used to describe the combined effects of heat conduction within the fluid (diffusion) and heat transference by bulk fluid flow streaming. The process of transport by fluid streaming is known as advection, but pure advection is a term that is generally associated only with mass transport in fluids, such as advection of pebbles in a river. In the case of heat transfer in fluids, where transport by advection in a fluid is always also accompanied by transport via heat diffusion (also known as heat conduction) the process of heat convection is understood to refer to the sum of heat transport by advection and diffusion/conduction.

Free, or natural, convection occurs when bulk fluid motions (streams and currents) are caused by buoyancy forces that result from density variations due to variations of temperature in the fluid. *Forced* convection is a term used when the streams and currents in the fluid are induced by external means—such as fans, stirrers, and pumps—creating an artificially induced convection current.

Convection-cooling

Convective cooling is sometimes described as Newton's law of cooling:

The rate of heat loss of a body is proportional to the temperature difference between the body and its surroundings.

However, by definition, the validity of Newton's law of cooling requires that the rate of heat loss from convection be a linear function of ("proportional to") the temperature difference that drives heat transfer, and in convective cooling this is sometimes not the case. In general, convection is not linearly dependent on temperature gradients, and in some cases is strongly nonlinear. In these cases, Newton's law does not apply.

Thermal radiation occurs through a vacuum or any transparent medium (solid or fluid). It is the transfer of energy by means of photons in electromagnetic waves governed by the same laws. Earth's radiation balance depends on the incoming and the outgoing thermal radiation, Earth's energy budget. Anthropogenic perturbations in the climate system are responsible for a positive radiative forcing which reduces the net longwave radiation loss to space.

Thermal radiation is energy emitted by matter as electromagnetic waves, due to the pool of thermal energy in all matter with a temperature above absolute zero. Thermal radiation propagates without the presence of matter through the vacuum of space.

Thermal radiation is a direct result of the random movements of atoms and molecules in matter. Since these atoms and molecules are composed of charged particles (protons and electrons), their movement results in the emission of electromagnetic radiation, which carries energy away from the surface.

Phase Transition

Phase transition or phase change, takes place in a thermodynamic system from one phase or state of matter to another one by heat transfer. Phase change examples are the melting of ice or the boiling of water. The Mason equation explains the growth of a water droplet based on the effects of heat transport on evaporation and condensation.

Types of phase transition occurring in the four fundamental states of matter, include:

- Solid - Deposition, freezing and solid to solid transformation.

- Gas - Boiling/evaporation, recombination/deionization, and sublimation.

- Liquid - Condensation and melting/fusion.

- Plasma - Ionization.

Lightning is a highly visible form of energy transfer and is an example of plasma present at Earth's surface. Typically, lightning discharges 30,000 amperes at up to 100 million volts, and emits light, radio waves, X-rays and even gamma rays. Plasma temperatures in lightning can approach 28,000 Kelvin (27,726.85 °C) (49,940.33 °F) and electron densities may exceed 10^{24} m^{-3}.

Boiling

Nucleate boiling of water.

The boiling point of a substance is the temperature at which the vapor pressure of the liquid equals the pressure surrounding the liquid and the liquid evaporates resulting in an abrupt change in vapor volume.

Saturation temperature means boiling point. The saturation temperature is the temperature for a corresponding saturation pressure at which a liquid boils into its vapor phase. The liquid can be said to be saturated with thermal energy. Any addition of thermal energy results in a phase transition.

At standard atmospheric pressure and low temperatures, no boiling occurs and the heat transfer rate is controlled by the usual single-phase mechanisms. As the surface temperature is increased, local boiling occurs and vapor bubbles nucleate, grow into the surrounding cooler fluid, and collapse. This is *sub-cooled nucleate boiling*, and is a very efficient heat transfer mechanism. At high bubble generation rates, the bubbles begin to interfere and the heat flux no longer increases rapidly with surface temperature (this is the departure from nucleate boiling, or DNB).

At similar standard atmospheric pressure and high temperatures, the hydrodynamically-quieter regime of film boiling is reached. Heat fluxes across the stable vapor layers are low, but rise slowly with temperature. Any contact between fluid and the surface that may be seen probably leads to the extremely rapid nucleation of a fresh vapor layer ("spontaneous nucleation"). At higher temperatures still, a maximum in the heat flux is reached (the critical heat flux, or CHF).

The Leidenfrost Effect demonstrates how nucleate boiling slows heat transfer due to gas bubbles on the heater's surface. As mentioned, gas-phase thermal conductivity is much lower than liquid-phase thermal conductivity, so the outcome is a kind of "gas thermal barrier".

Condensation

Condensation occurs when a vapor is cooled and changes its phase to a liquid. During condensation, the latent heat of vaporization must be released. The amount of the heat is the same as that absorbed during vaporization at the same fluid pressure.

There are several types of condensation:

- Homogeneous condensation, as during a formation of fog.

- Condensation in direct contact with subcooled liquid.

- Condensation on direct contact with a cooling wall of a heat exchanger: This is the most common mode used in industry:

    o Filmwise condensation is when a liquid film is formed on the subcooled surface, and usually occurs when the liquid wets the surface.

    o Dropwise condensation is when liquid drops are formed on the subcooled surface, and usually occurs when the liquid does not wet the surface.

    Dropwise condensation is difficult to sustain reliably; therefore, industrial equipment is normally designed to operate in filmwise condensation mode.

Melting

Melting is a physical process that results in the phase transition of a substance from a solid to a liquid. The internal energy of a substance is increased, typically by the application of heat or pressure, resulting in a rise of its temperature to the melting point, at which the ordering of ionic or molecular entities in the solid breaks down to a less ordered state and the solid liquefies. An object that has melted completely is molten. Substances in the molten state generally have reduced viscosity with elevated temperature; an exception to this maxim is the element sulfur, whose viscosity increases to a point due to polymerization and then decreases with higher temperatures in its molten state.

Ice melting

Modeling Approaches

Heat transfer can be modeled in the following ways.

Climate Models

Climate models study the radiant heat transfer by using quantitative methods to simulate the interactions of the atmosphere, oceans, land surface, and ice.

Heat Equation

The heat equation is an important partial differential equation that describes the distribution of heat (or variation in temperature) in a given region over time. In some cases, exact solutions of the equation are available; in other cases the equation must be solved numerically using computational methods.

Lumped System Analysis

Lumped system analysis often reduces the complexity of the equations to one first-order linear differential equation, in which case heating and cooling are described by a simple exponential solution, often referred to as Newton's law of cooling.

System analysis by the lumped capacitance model is a common approximation in transient conduction that may be used whenever heat conduction within an object is much faster than heat conduction across the boundary of the object. This is a method of approximation that reduces one aspect of the transient conduction system—that within the object—to an equivalent steady state system. That is, the method assumes that the temperature within the object is completely uniform, although its value may be changing in time.

In this method, the ratio of the conductive heat resistance within the object to the convective heat transfer resistance across the object's boundary, known as the *Biot number*, is calculated. For small Biot numbers, the approximation of *spatially uniform temperature within the object* can be used: it can be presumed that heat transferred into the object has time to uniformly distribute itself, due to the lower resistance to doing so, as compared with the resistance to heat entering the object.

Engineering

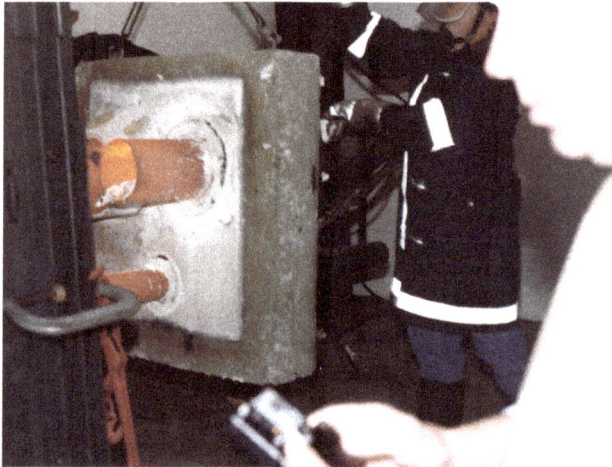

Heat exposure as part of a fire test for firestop products

Heat transfer has broad application to the functioning of numerous devices and systems. Heat-transfer principles may be used to preserve, increase, or decrease temperature in a wide variety of circumstances.Heat transfer methods are used in numerous disciplines, such as automotive engineering, thermal management of electronic devices and systems, climate control, insulation, materials processing, and power station engineering.

Insulation, Radiance and Resistance

Thermal insulators are materials specifically designed to reduce the flow of heat by limiting conduction, convection, or both. Thermal resistance is a heat property and the measurement by which an object or material resists to heat flow (heat per time unit or thermal resistance) to temperature difference.

Radiance or spectral radiance are measures of the quantity of radiation that passes through or is emitted. Radiant barriers are materials that reflect radiation, and therefore reduce the flow of heat from radiation sources. Good insulators are not necessarily good radiant barriers, and vice versa. Metal, for instance, is an excellent reflector and a poor insulator.

The effectiveness of a radiant barrier is indicated by its reflectivity, which is the fraction of radiation reflected. A material with a high reflectivity (at a given wavelength) has a low emissivity (at that same wavelength), and vice versa. At any specific wavelength, reflectivity = 1 - emissivity. An ideal radiant barrier would have a reflectivity of 1, and would therefore reflect 100 percent of incoming radiation. Vacuum flasks, or Dewars, are silvered to approach this ideal. In the vacuum of space, satellites use multi-layer insulation, which consists of many layers of aluminized (shiny) Mylar to greatly reduce radiation heat transfer and control satellite temperature.

Devices

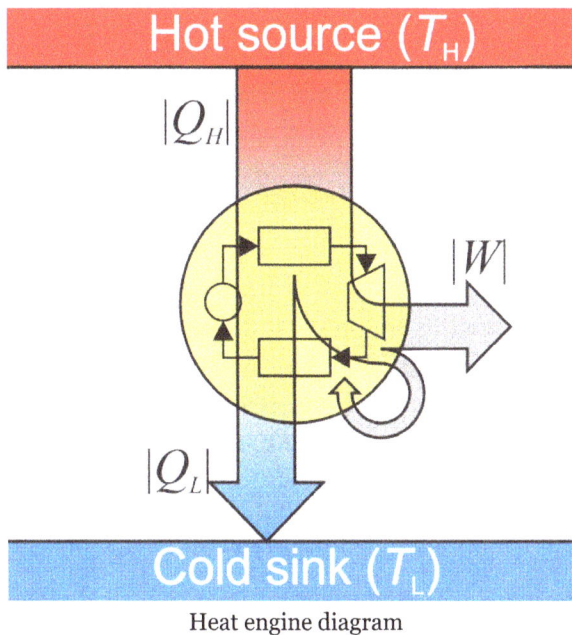

Heat engine diagram

- Heat engine is a system that performs the conversion of heat or thermal energy to mechanical energy which can then be used to do mechanical work.

- Thermocouple is a temperature-measuring device and widely used type of temperature sensor for measurement and control, and can also be used to convert heat into electric power.

- Thermoelectric cooler is a solid state electronic device that pumps (transfers) heat from one side of the device to the other when electric current is passed through it. It is based on the Peltier effect.

- Thermal diode or thermal rectifier is a device that causes heat to flow preferentially in one direction.

Heat Exchangers

A heat exchanger is used for more efficient heat transfer or to dissipate heat. Heat exchangers are widely used in refrigeration, air conditioning, space heating, power generation, and chemical processing. One common example of a heat exchanger is a car's radiator, in which the hot coolant fluid is cooled by the flow of air over the radiator's surface.

Common types of heat exchanger flows include parallel flow, counter flow, and cross flow. In parallel flow, both fluids move in the same direction while transferring heat; in counter flow, the fluids move in opposite directions; and in cross flow, the fluids move at right angles to each other. Common constructions for heat exchanger include shell and tube, double pipe, extruded finned pipe, spiral fin pipe, u-tube, and stacked plate.

A heat sink is a component that transfers heat generated within a solid material to a fluid medium, such as air or a liquid. Examples of heat sinks are the heat exchangers used in refrigeration and air conditioning systems or the radiator in a car. A heat pipe is another heat-transfer device that combines thermal conductivity and phase transition to efficiently transfer heat between two solid interfaces.

Examples

Architecture

Efficient energy use is the goal to reduce the amount of energy required in heating or cooling. In architecture, condensation and air currents can cause cosmetic or structural damage. An energy audit can help to assess the implementation of recommended corrective procedures. For instance, insulation improvements, air sealing of structural leaks or the addition of energy-efficient windows and doors.

- Smart meter is a device that records electric energy consumption in intervals.

- Thermal transmittance is the rate of transfer of heat through a structure divided by the difference in temperature across the structure. It is expressed in watts per square meter per kelvin, or W/m^2K. Well-insulated parts of a building have a low thermal transmittance, whereas poorly-insulated parts of a building have a high thermal transmittance.

- Thermostat is a device to monitor and control temperature.

Climate Engineering

Climate engineering consists of carbon dioxide removal and solar radiation management. Since the amount of carbon dioxide determines the radiative balance of Earth atmosphere, carbon dioxide removal techniques can be applied to reduce the radiative forcing. Solar radiation management is the attempt to absorb less solar radiation to offset the effects of greenhouse gases.

An example application in climate engineering includes the creation of Biochar through the pyrolysis process. Thus, storing greenhouse gases in carbon reduces the radiative forcing capacity in the atmosphere, causing more long-wave (infrared) radiation out to Space.

Greenhouse Effect

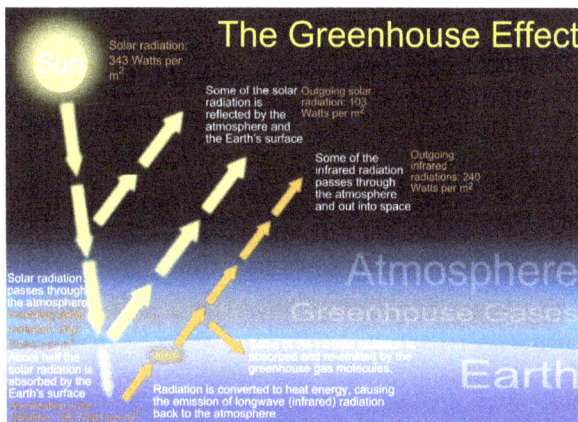

A representation of the exchanges of energy between the source (the Sun), the Earth's surface, the Earth's atmosphere, and the ultimate sink outer space. The ability of the atmosphere to capture and recycle energy emitted by the Earth surface is the defining characteristic of the greenhouse effect.

The greenhouse effect is a process by which thermal radiation from a planetary surface is absorbed by atmospheric greenhouse gases, and is re-radiated in all directions. Since part of this re-radiation is back towards the surface and the lower atmosphere, it results in an elevation of the average surface temperature above what it would be in the absence of the gases.

Heat Transfer in the Human Body

The principles of heat transfer in engineering systems can be applied to the human body in order to determine how the body transfers heat. Heat is produced in the body by the continuous metabolism of nutrients which provides energy for the systems of the body. The human body must maintain a consistent internal temperature in order to maintain healthy bodily functions. Therefore, excess heat must be dissipated from the body to keep it from overheating. When a person engages in elevated levels of physical activity, the body requires additional fuel which increases the metabolic rate and the rate of heat production. The body must then use additional methods to remove the additional heat produced in order to keep the internal temperature at a healthy level.

Heat transfer by convection is driven by the movement of fluids over the surface of the body. This convective fluid can be either a liquid or a gas. For heat transfer from the outer surface of the body, the convection mechanism is dependent on the surface area of the body, the velocity of the air, and the temperature gradient between the surface of the skin and the ambient air. The normal temperature of the body is approximately 37 °C. Heat transfer occurs more readily when the temperature of the surroundings is significantly less than the normal body temperature. This concept explains why a person feels "cold" when not enough covering is worn when exposed to a cold environment. Clothing can be considered an insulator which provides thermal resistance to heat flow over the covered portion of the body. This thermal resistance causes the temperature on the surface of the clothing to be less than the temperature on the surface of the skin. This smaller temperature gradient between the surface temperature and the ambient temperature will cause a lower rate of heat transfer than if the skin were not covered.

In order to ensure that one portion of the body is not significantly hotter than another portion, heat must be distributed evenly through the bodily tissues. Blood flowing through blood vessels acts as a convective fluid and helps to prevent any buildup of excess heat inside the tissues of the body. This flow of blood through the vessels can be modeled as pipe flow in an engineering system. The heat carried by the blood is determined by the temperature of the surrounding tissue, the diameter of the blood vessel, the thickness of the fluid, velocity of the flow, and the heat transfer coefficient of the blood. The velocity, blood vessel diameter, and the fluid thickness can all be related with the Reynolds Number, a dimensionless number used in fluid mechanics to characterize the flow of fluids.

Latent heat loss, also known as evaporative heat loss, accounts for a large fraction of heat loss from the body. When the core temperature of the body increases, the body triggers sweat glands in the skin to bring additional moisture to the surface of the skin. The liquid is then transformed into vapor which removes heat from the surface of the body. The rate of evaporation heat loss is directly related to the vapor pressure at the skin surface and the amount of moisture present on the skin. Therefore, the maximum of heat transfer will occur when the skin is completely wet. The body continuously loses water by evaporation but the most significant amount of heat loss occurs during periods of increased physical activity.

Cooling Techniques

Evaporative Cooling

A traditional air cooler in Mirzapur, Uttar Pradesh, India

Evaporative cooling happens when water vapor is added to the surrounding air. The energy needed to evaporate the water is taken from the air in the form of sensible heat and converted into latent heat, while the air remains at a constant enthalpy. Latent heat describes the amount of heat that is needed to evaporate the liquid; this heat comes from the liquid itself and the surrounding gas and surfaces. The greater the difference between the two temperatures, the greater the evaporative cooling effect. When the temperatures are the same, no net evaporation of water in air occurs; thus, there is no cooling effect.

Laser Cooling

In Quantum Physics laser cooling is used to achieve temperatures of near absolute zero (−273.15 °C, −459.67 °F) of atomic and molecular samples, to observe unique quantum effects that can only occur at this heat level.

- Doppler cooling is the most common method of laser cooling.

- Sympathetic cooling is a process in which particles of one type cool particles of another type. Typically, atomic ions that can be directly laser-cooled are used to cool nearby ions or atoms. This technique allows cooling of ions and atoms that cannot be laser cooled directly.

Magnetic Cooling

Magnetic evaporative cooling is a process for lowering the temperature of a group of atoms, after pre-cooled by methods such as laser cooling. Magnetic refrigeration cools below 0.3K, by making use of the magnetocaloric effect.

Radiative Cooling

Radiative cooling is the process by which a body loses heat by radiation. Outgoing energy is an important effect in the Earth's energy budget. In the case of the Earth-atmosphere system, it refers to the process by which long-wave (infrared) radiation is emitted to balance the absorption of short-wave (visible) energy from the Sun. Convective transport of heat and evaporative transport of latent heat both remove heat from the surface and redistribute it in the atmosphere.

Thermal Energy Storage

Thermal energy storage refers to technologies used to collect and store energy for later use. They can be employed to balance energy demand between day and nighttime. The thermal reservoir may be maintained at a temperature above (hotter) or below (colder) than that of the ambient environment. Applications include later use in space heating, domestic or process hot water, or to generate electricity.

Mass Transfer

Mass transfer is the net movement of mass from one location, usually meaning stream, phase, fraction or component, to another. Mass transfer occurs in many processes, such as absorption, evaporation, adsorption, drying, precipitation, membrane filtration, and distillation. Mass transfer is used by different scientific disciplines for different processes and mechanisms. The phrase is commonly used in engineering for physical processes that involve diffusive and convective transport of chemical species within physical systems.

Some common examples of mass transfer processes are the evaporation of water from a pond to the atmosphere, the purification of blood in the kidneys and liver, and the distillation of alcohol. In industrial processes, mass transfer operations include separation of chemical components in distillation columns, absorbers such as scrubbers or stripping, adsorbers such as activated carbon beds, and liquid-liquid extraction. Mass transfer is often coupled to additional transport processes, for instance in industrial cooling towers. These towers couple heat transfer to mass transfer by allowing hot water to flow in contact with hotter air and evaporate as it absorbs heat from the air.

Astrophysics

In astrophysics, mass transfer is the process by which matter gravitationally bound to a body, usually a star, fills its Roche lobe and becomes gravitationally bound to a second body, usually a compact object (white dwarf, neutron star or black hole), and is eventually accreted onto it. It is a common phenomenon in binary systems, and may play an important role in some types of supernovae and pulsars.

Chemical Engineering

Mass transfer finds extensive application in chemical engineering problems. It is used in reaction engineering, separations engineering, heat transfer engineering, and many other sub-disciplines of chemical engineering.

The driving force for mass transfer is typically a difference in chemical potential, when it can be defined, though other thermodynamic gradients may couple to the flow of mass and drive it as well. A chemical species moves from areas of high chemical potential to areas of low chemical potential. Thus, the maximum theoretical extent of a given mass transfer is typically determined by the point at which the chemical potential is uniform. For single phase-systems, this usually translates to uniform concentration throughout the phase, while for multiphase systems chemical species will often prefer one phase over the others and reach a uniform chemical potential only when most of the chemical species has been absorbed into the preferred phase, as in liquid-liquid extraction.

While thermodynamic equilibrium determines the theoretical extent of a given mass transfer operation, the actual rate of mass transfer will depend on additional factors including the flow patterns within the system and the diffusivities of the species in each phase. This rate can be quantified through the calculation and application of mass transfer coefficients for an overall process. These mass transfer coefficients are typically published in terms of dimensionless numbers, often including Péclet numbers, Reynolds numbers, Sherwood numbers and Schmidt numbers, among others.

Analogies Between Heat, Mass, and Momentum Transfer

There are notable similarities in the commonly used approximate differential equations for momentum, heat, and mass transfer. The molecular transfer equations of Newton's law for fluid momentum at low Reynolds number (Stokes flow), Fourier's law for heat, and Fick's law for mass are very similar, since they are all linear approximations to transport of conserved quantities in a flow field. At higher Reynolds number, the analogy between mass and heat transfer and momentum transfer becomes less useful due to the nonlinearity of the Navier-Stokes equation (or more fundamentally, the general momentum conservation equation), but the analogy between heat and mass transfer remains good. A great deal of effort has been devoted to developing analogies among these three transport processes so as to allow prediction of one from any of the others.

References

- Robert F. Speyer (2012). Thermal Analysis of Materials. Materials Engineering. Marcel Dekker, Inc. p. 2. ISBN 0-8247-8963-6.

- Law, C.K. (2006). Combustion Physics. Cambridge, UK: Cambridge University Press. ISBN 9780521154215.

- Paul A., Tipler; Gene Mosca (2008). Physics for Scientists and Engineers, Volume 1 (6th ed.). New York, NY: Worth Publishers. pp. 666–670. ISBN 1-4292-0132-0.

- Lienhard, John H.,V; Lienhard, John H., V (2008). A Heat Transfer Textbook (3rd ed.). Cambridge, Massachusetts: Phlogiston Press. ISBN 978-0-9713835-3-1. OCLC 230956959.

- Welty, James R.; Wicks, Charles E.; Wilson, Robert Elliott (1976). Fundamentals of momentum, heat, and mass transfer (2 ed.). New York: Wiley. ISBN 978-0-471-93354-0. OCLC 2213384.

- Faghri, Amir; Zhang, Yuwen; Howell, John (2010). Advanced Heat and Mass Transfer. Columbia, MO: Global Digital Press. ISBN 978-0-9842760-0-4.

- Abbott, J.M. Smith, H.C. Van Ness, M.M. (2005). Introduction to chemical engineering thermodynamics (7th ed.). Boston ; Montreal: McGraw-Hill. ISBN 0-07-310445-0.

- Çengel, Yunus (2003). Heat Transfer: a practical approach. McGraw-Hill series in mechanical engineering. (2nd ed.). Boston: McGraw-Hill. ISBN 978-0-07-245893-0. OCLC 300472921. Retrieved 2009-04-20.

- Geankoplis, Christie John (2003). Transport processes and separation process principles : (includes unit operations) (4th ed.). Upper Saddle River, NJ: Prentice Hall Professional Technical Reference. ISBN 0-13-101367-X.

- David.E. Goldberg (1988). 3,000 Solved Problems in Chemistry (1st ed.). McGraw-Hill. ISBN 0-07-023684-4. Section 17.43, page 321

- Louis Theodore, R. Ryan Dupont and Kumar Ganesan (Editors) (1999). Pollution Prevention: The Waste Management Approach to the 21st Century. CRC Press. ISBN 1-56670-495-2. Section 27, page 15

- Taylor, R.A. (2012). "Socioeconomic impacts of heat transfer research". International Communications in Heat and Mass Transfer. 39 (10): 1467–1473.

- "EnergySavers: Tips on Saving Money & Energy at Home" (PDF). U.S. Department of Energy. Retrieved March 2, 2012.

- New Jersey Institute of Technology, Chemical Engineering Dept. "B.S. Chemical Engineering". NJIT. Retrieved 9 April 2011.

- C.Michael Hogan (2011) Sulfur, Encyclopedia of Earth, eds. A.Jorgensen and C.J.Cleveland, National Council for Science and the environment, Washington DC

Processes of Thermal Engineering

The process that causes the release of energy in any system to its surrounding is known as exothermic process. This release is usually in the form of heat. The exact opposite of exothermic process is endothermic process, while exothermic process releases energy into the environment, endothermic process is the process of absorbing energy from its environment. This section helps the readers in understanding the process related to thermal engineering.

Exothermic Process

Explosions are some of the most violent exothermic reactions.

In thermodynamics, the term exothermic process (exo- : "outside") describes a process or reaction that releases energy from the system to its surroundings, usually in the form of heat, but also in a form of light (e.g. a spark, flame, or flash), electricity (e.g. a battery), or sound (e.g. explosion heard when burning hydrogen). The term *exothermic* was first coined by Marcellin Berthelot. The opposite of an exothermic process is an endothermic process, one that absorbs energy in the form of heat.

The concept is frequently applied in the physical sciences to chemical reactions, where as in chemical bond energy that will be converted to thermal energy (heat).

Exothermic (and endothermic) describe two types of chemical reactions or systems found in nature, as follows.

Simply stated, after an exothermic reaction, more energy has been released to the surroundings than was absorbed to initiate and maintain the reaction. An example would be the burning of a candle, wherein the sum of calories produced by combustion (found by looking at radiant heating of the surroundings and visible light produced, including increase in temperature of the fuel (wax) itself, which with oxygen, have become hot CO_2 and water vapor,) exceeds the number of calories absorbed initially in lighting the flame and in the flame maintaining itself. (i.e. some energy produced by combustion is reabsorbed and used in melting, then vaporizing the wax, etc. but is (far) outstripped by the energy produced in breaking carbon-hydrogen bonds and combination of oxygen with the resulting carbon and hydrogen).

On the other hand, in an endothermic reaction or system, energy is taken from the surroundings in the course of the reaction. An example of an endothermic reaction is a first aid cold pack, in which the reaction of two chemicals, or dissolving of one in another, requires calories from the surroundings, and the reaction cools the pouch and surroundings by absorbing heat from them. An endothermic system is seen in the production of wood: trees absorb radiant energy, from the sun, use it in endothermic reactions such as taking apart CO_2 and H_2O and combining the carbon and hydrogen generated to produce cellulose and other organic chemicals. These products, in the form of wood, say, may later be burned in a fireplace, exothermically, producing CO_2 and water, and releasing energy in the form of heat and light to their surroundings, e.g., to a home's interior and chimney gasses.

Overview

Exothermic refers to a transformation in which a system releases energy (heat) to the surroundings, expressed by

$$Q < 0.$$

When the transformation occurs at constant pressure, one has for the enthalpy

$$\Delta H < 0,$$

and constant volume, one has for the internal energy

$$\Delta U < 0.$$

In an adiabatic system (i.e. a system that does not exchange heat with the surroundings), an exothermic process results in an increase in temperature of the system.

In exothermic chemical reactions, the heat that is released by the reaction takes the form of electromagnetic energy. The transition of electrons from one quantum energy level to another causes light to be released. This light is equivalent in energy to the stabilization energy of the energy for the chemical reaction, i.e. the bond energy. This light that is released can be absorbed by other molecules in solution to give rise to molecular vibrations or rotations, which gives rise to the classical understanding of heat. In contrast, when endothermic reactions occur, energy is absorbed to place an electron in a higher energy state, such that the electron can associate with another atom to form a chemical complex. Net energy is absorbed by an endothermic reaction. In an exothermic reaction, the energy needed to start the reaction is less than the energy that is subsequently released, so there is a net release of energy. This is the physical understanding of exothermic and endothermic reactions.

Examples

An exothermic thermite reaction using iron(III) oxide. The sparks flying outwards are globules of molten iron trailing smoke in their wake.

Some examples of exothermic processes are:

- Combustion of fuels such as wood, coal and oil petroleum

- Thermite reaction

- Reaction of alkali metals and other highly electropositive metals with water

- Condensation of rain from water vapor

- Mixing water and strong acids or strong bases

- Mixing acids and bases

- Dehydration of carbohydrates by sulfuric acid

- The setting of cement and concrete

- Some polymerisation reactions such as the setting of epoxy resin

- Reaction of most metals with halogens or oxygen

- Nuclear fusion in hydrogen bombs and in stellar cores (to iron)

- Nuclear fission of heavy elements

Implications for Chemical Reactions

Chemical exothermic reactions are generally more spontaneous than their counterparts, endothermic reactions.

In a thermochemical reaction that is exothermic, the heat may be listed among the products of the reaction.

Contrast Between Thermodynamic and Biological Terminology

Because of historical accident, students encounter a source of possible confusion between the terminology of physics and biology. Whereas the *thermodynamic* terms "exothermic" and "endothermic" respectively refer to processes that give out heat energy and processes that absorb heat energy, in *biology* the sense is effectively inverted. The metabolic terms "ectothermic" and "endothermic" respectively refer to organisms that rely largely on external heat to achieve a full working temperature, and to organisms that produce heat from within as a major factor in controlling their bodily temperature.

Endothermic Process

The term endothermic process describes a process or reaction in which the system absorbs energy from its surroundings; usually, but not always, in the form of heat. The intended sense is that of a reaction that depends on absorbing heat if it is to proceed. The opposite of an endothermic process is an exothermic process, one that releases, "gives out" energy in the form of (usually, but not always) heat. Thus in each term (endothermic & exothermic) the prefix refers to where heat goes as the reaction occurs, though in reality it only refers to where the energy goes, without necessarily being in the form of heat.

The concept is frequently applied in physical sciences to, for example, chemical reactions, where thermal energy (heat) is converted to chemical bond energy.

Endothermic (and exothermic) analysis only accounts for the enthalpy change (ΔH) of a reaction. The full energy analysis of a reaction is the Gibbs free energy (ΔG), which includes an entropy (ΔS) and temperature term in addition to the enthalpy. A reaction will be a spontaneous process at a certain temperature if the products have a lower Gibbs free energy (an exergonic reaction) even if the enthalpy of the products is higher. Entropy and enthalpy are different terms, so the change in entropic energy can overcome an opposite change in enthalpic energy and make an endothermic reaction favorable.

Examples

- Photosynthesis

- Melting ice

- Evaporating liquid water

- Sublimation of carbon dioxide (dry ice)

- Cracking of alkanes

- Thermal decomposition reactions

- Electrolytic decomposition of sodium chloride into sodium hydroxide and hydrogen chloride

- Dissolving ammonium chloride in water

- Nucleosynthesis of elements heavier than nickel in stellar cores

- High-energy neutrons can produce tritium from lithium-7 in an endothermic reaction, consuming 2.466 MeV. This was discovered when the 1954 Castle Bravo nuclear test produced an unexpectedly high yield.

- Nuclear fusion of elements heavier than iron in supernovae

References

- Austin, Patrick (January 1996). "Tritium: The environmental, health, budgetary, and strategic effects of the Department of Energy's decision to produce tritium". Institute for Energy and Environmental Research. Retrieved 2010-09-15.

4

Understanding Thermodynamics

In order to develop an in-depth understanding of thermodynamics, it is very important to understand certain topics. Some of these topics are thermodynamics, internal energy, thermodynamic operation, equilibrium thermodynamics and state function. The topics discussed in the chapter are of great importance to broaden the existing knowledge on thermal engineering.

Thermodynamics

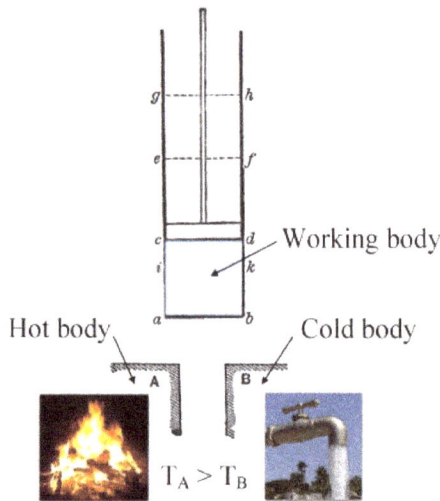

Annotated color version of the original 1824 Carnot heat engine showing the hot body (boiler), working body (system, steam), and cold body (water), the letters labeled according to the stopping points in Carnot cycle.

Thermodynamics is a branch of science concerned with heat and temperature and their relation to energy and work. The behavior of these quantities is governed by the four laws of thermodynamics, irrespective of the composition or specific properties of the material or system in question. The laws of thermodynamics are explained in terms of microscopic constituents by statistical mechanics. Thermodynamics applies to a wide variety of topics in science and engineering, especially physical chemistry, chemical engineering and mechanical engineering.

Historically, thermodynamics developed out of a desire to increase the efficiency of early steam engines, particularly through the work of French physicist Nicolas Léonard

Sadi Carnot (1824) who believed that engine efficiency was the key that could help France win the Napoleonic Wars. Scottish physicist Lord Kelvin was the first to formulate a concise definition of thermodynamics in 1854:

Thermo-dynamics is the subject of the relation of heat to forces acting between contiguous parts of bodies, and the relation of heat to electrical agency.

The initial application of thermodynamics to mechanical heat engines was extended early on to the study of chemical systems. Chemical thermodynamics studies the nature of the role of entropy in the process of chemical reactions and provided the bulk of expansion and knowledge of the field. Other formulations of thermodynamics emerged in the following decades. Statistical thermodynamics, or statistical mechanics, concerned itself with statistical predictions of the collective motion of particles from their microscopic behavior. In 1909, Constantin Carathéodory presented a purely mathematical approach to the field in his axiomatic formulation of thermodynamics, a description often referred to as *geometrical thermodynamics*.

Introduction

The starting point for most considerations of thermodynamic systems are the laws of thermodynamics, four principles that form an axiomatic basis. The first law specifies that energy can be exchanged between physical systems as heat and work. The second law defines the existence of the quantity entropy, a quantification of the state of order of a system that expresses the notion of useful work that can be performed.

In thermodynamics, interactions between large ensembles of objects are studied and categorized. Central to this are the concepts of *system* and *surroundings*. A system is composed of particles, whose average motions define its properties, which in turn are related to one another through equations of state. Properties can be combined to express internal energy and thermodynamic potentials, which are useful for determining conditions for equilibrium and spontaneous processes.

With these tools, thermodynamics can be used to describe how systems respond to changes in their environment. This can be applied to a wide variety of topics in science and engineering, such as engines, phase transitions, chemical reactions, transport phenomena, and even black holes. The results of thermodynamics are essential for other fields of physics and for chemistry, chemical engineering, aerospace engineering, mechanical engineering, cell biology, biomedical engineering, materials science, and economics, to name a few.

This article is focused mainly on classical thermodynamics which primarily studies systems in thermodynamic equilibrium. Non-equilibrium thermodynamics is often treated as an extension of the classical treatment, but statistical mechanics has brought many advances of the field.

École Polytechnique	Glasgow school	Berlin school	Edinburgh school
Sadi Carnot (1796-1832)	William Thomson (1824-1907)	Rudolf Clausius (1822-1888)	James Maxwell (1831-1879)
Vienna school	Gibbsian school	Dresden school	Dutch school
Ludwig Boltzmann (1844-1906)	Willard Gibbs (1839-1903)	Gustav Zeuner (1828-1907)	Johannes der Waals (1837-1923)

The thermodynamicists representative of the original eight founding schools of thermodynamics. The schools with the most-lasting effect in founding the modern versions of thermodynamics are the Berlin school, particularly as established in Rudolf Clausius's 1865 textbook *The Mechanical Theory of Heat*, the Vienna school, with the statistical mechanics of Ludwig Boltzmann, and the Gibbsian school at Yale University, American engineer Willard Gibbs' 1876 *On the Equilibrium of Heterogeneous Substances* launching chemical thermodynamics.

History

The history of thermodynamics as a scientific discipline generally begins with Otto von Guericke who, in 1650, built and designed the world's first vacuum pump and demonstrated a vacuum using his Magdeburg hemispheres. Guericke was driven to make a vacuum in order to disprove Aristotle's long-held supposition that 'nature abhors a vacuum'. Shortly after Guericke, the English physicist and chemist Robert Boyle had learned of Guericke's designs and, in 1656, in coordination with English scientist Robert Hooke, built an air pump. Using this pump, Boyle and Hooke noticed a correlation between pressure, temperature, and volume. In time, Boyle's Law was formulated, which states that pressure and volume are inversely proportional. Then, in 1679, based on these concepts, an associate of Boyle's named Denis Papin built a steam digester, which was a closed vessel with a tightly fitting lid that confined steam until a high pressure was generated.

Later designs implemented a steam release valve that kept the machine from exploding. By watching the valve rhythmically move up and down, Papin conceived of the idea of a piston and a cylinder engine. He did not, however, follow through with his design. Nevertheless, in 1697, based on Papin's designs, engineer Thomas Savery built the first engine, followed by Thomas Newcomen in 1712. Although these early engines were crude and inefficient, they attracted the attention of the leading scientists of the time.

The fundamental concepts of heat capacity and latent heat, which were necessary for the development of thermodynamics, were developed by Professor Joseph Black at the University of Glasgow, where James Watt was employed as an instrument maker. Black and Watt performed experiments together, but it was Watt who conceived the idea of the external condenser which resulted in a large increase in steam engine efficiency. Drawing on all the previous work led Sadi Carnot, the "father of thermodynamics", to publish *Reflections on the Motive Power of Fire* (1824), a discourse on heat, power, energy and engine efficiency. The paper outlined the basic energetic relations between the Carnot engine, the Carnot cycle, and motive power. It marked the start of thermodynamics as a modern science.

The first thermodynamic textbook was written in 1859 by William Rankine, originally trained as a physicist and a civil and mechanical engineering professor at the University of Glasgow. The first and second laws of thermodynamics emerged simultaneously in the 1850s, primarily out of the works of William Rankine, Rudolf Clausius, and William Thomson (Lord Kelvin).

The foundations of statistical thermodynamics were set out by physicists such as James Clerk Maxwell, Ludwig Boltzmann, Max Planck, Rudolf Clausius and J. Willard Gibbs.

During the years 1873-76 the American mathematical physicist Josiah Willard Gibbs published a series of three papers, the most famous being *On the Equilibrium of Heterogeneous Substances*, in which he showed how thermodynamic processes, including chemical reactions, could be graphically analyzed, by studying the energy, entropy, volume, temperature and pressure of the thermodynamic system in such a manner, one can determine if a process would occur spontaneously. Also Pierre Duhem in the 19th century wrote about chemical thermodynamics. During the early 20th century, chemists such as Gilbert N. Lewis, Merle Randall, and E. A. Guggenheim applied the mathematical methods of Gibbs to the analysis of chemical processes.

Etymology

The etymology of *thermodynamics* has an intricate history. It was first spelled in a hyphenated form as an adjective (*thermo-dynamic*) and from 1854 to 1868 as the noun *thermo-dynamics* to represent the science of generalized heat engines.

Pierre Perrot claims that the term *thermodynamics* was coined by James Joule in 1858 to designate the science of relations between heat and power, however, Joule never used that term, but used instead the term *perfect thermo-dynamic engine* in reference to Thomson's 1849 phraseology.

By 1858, *thermo-dynamics*, as a functional term, was used in William Thomson's paper *An Account of Carnot's Theory of the Motive Power of Heat*.

Branches of Description

The study of thermodynamical systems has developed into several related branches, each using a different fundamental model as a theoretical or experimental basis, or applying the principles to varying types of systems.

Classical Thermodynamics

Classical thermodynamics is the description of the states of thermodynamical systems at near-equilibrium, using macroscopic, empirical properties directly measurable in the laboratory. It is used to model exchanges of energy, work and heat based on the laws of thermodynamics. The qualifier *classical* reflects the fact that it represents the descriptive level in terms of macroscopic empirical parameters that can be measured in the laboratory, that was the first level of understanding in the 19th century. A microscopic interpretation of these concepts was provided by the development of statistical mechanics.

Statistical Mechanics

Statistical mechanics, also called statistical thermodynamics, emerged with the development of atomic and molecular theories in the late 19th century and early 20th century, supplementing thermodynamics with an interpretation of the microscopic interactions between individual particles or quantum-mechanical states. This field relates the microscopic properties of individual atoms and molecules to the macroscopic, bulk properties of materials that can be observed on the human scale, thereby explaining thermodynamics as a natural result of statistics, classical mechanics, and quantum theory at the microscopic level.

Chemical Thermodynamics

Chemical thermodynamics is the study of the interrelation of energy with chemical reactions or with a physical change of state within the confines of the laws of thermodynamics.

Treatment of Equilibrium

Equilibrium thermodynamics is the systematic study of transformations of matter and energy in systems as they approach equilibrium. The word equilibrium implies a state of balance. In an equilibrium state there are no unbalanced potentials, or driving forces, within the system. A central aim in equilibrium thermodynamics is: given a system in a well-defined initial state, subject to accurately specified constraints, to calculate what the state of the system will be once it has reached equilibrium.

Non-equilibrium thermodynamics is a branch of thermodynamics that deals with systems that are not in thermodynamic equilibrium. Most systems found in nature are not in thermodynamic equilibrium because they are not in stationary states, and are continuously and discontinuously subject to flux of matter and energy to and from other systems. The thermodynamic study of non-equilibrium systems requires more general concepts than are dealt with by equilibrium thermodynamics. Many natural systems still today remain beyond the scope of currently known macroscopic thermodynamic methods.

Laws of Thermodynamics

Thermodynamics is principally based on a set of four laws which are universally valid when applied to systems that fall within the constraints implied by each. In the various theoretical descriptions of thermodynamics these laws may be expressed in seemingly differing forms, but the most prominent formulations are the following:

- Zeroth law of thermodynamics: *If two systems are each in thermal equilibrium with a third, they are also in thermal equilibrium with each other.*

This statement implies that thermal equilibrium is an equivalence relation on the set of thermodynamic systems under consideration. Systems are said to be in equilibrium if the small, random exchanges between them (e.g. Brownian motion) do not lead to a net change in energy. This law is tacitly assumed in every measurement of temperature. Thus, if one seeks to decide if two bodies are at the same temperature, it is not necessary to bring them into contact and measure any changes of their observable properties in time. The law provides an empirical definition of temperature and justification for the construction of practical thermometers.

The zeroth law was not initially recognized as a law, as its basis in thermodynamical equilibrium was implied in the other laws. The first, second, and third laws had been explicitly stated prior and found common acceptance in the physics community. Once the importance of the zeroth law for the definition of temperature was realized, it was impracticable to renumber the other laws, hence it was numbered the *zeroth law*.

- First law of thermodynamics: *The internal energy of an isolated system is constant.*

The first law of thermodynamics is an expression of the principle of conservation of energy. It states that energy can be transformed (changed from one form to another), but cannot be created or destroyed.

The first law is usually formulated by saying that the change in the internal energy of a closed thermodynamic system is equal to the difference between the heat supplied to the system and the amount of work done by the system on its surroundings. It is important to note that internal energy is a state of the system whereas heat and work modify the state of the system. In other words, a specific internal energy of a system may be achieved by any combination of heat and work; the manner by which a system achieves a specific internal energy is path independent.

- Second law of thermodynamics: *Heat cannot spontaneously flow from a colder location to a hotter location.*

The second law of thermodynamics is an expression of the universal principle of decay observable in nature. The second law is an observation of the fact that over time, differences in temperature, pressure, and chemical potential tend to even out in a physical system that is isolated from the outside world. Entropy is a measure of how much this process has progressed. The entropy of an isolated system which is not in equilibrium will tend to increase over time, approaching a maximum value at equilibrium.

In classical thermodynamics, the second law is a basic postulate applicable to any system involving heat energy transfer; in statistical thermodynamics, the second law is a consequence of the assumed randomness of molecular chaos. There are many versions of the second law, but they all have the same effect, which is to explain the phenomenon of irreversibility in nature.

- Third law of thermodynamics: *As a system approaches absolute zero, all processes cease and the entropy of the system approaches a minimum value.*

The third law of thermodynamics is a statistical law of nature regarding entropy and the impossibility of reaching absolute zero of temperature. This law provides an absolute reference point for the determination of entropy. The entropy determined relative to this point is the absolute entropy. Alternate definitions are, "the entropy of all systems and of all states of a system is smallest at absolute zero," or equivalently "it is impossible to reach the absolute zero of temperature by any finite number of processes".

Absolute zero, at which all activity would stop if it were possible to happen, is −273.15 °C (degrees Celsius), or −459.67 °F (degrees Fahrenheit) or 0 K (kelvin).

System Models

An important concept in thermodynamics is the thermodynamic system, a precisely defined region of the universe under study. Everything in the universe except the system is known as the *surroundings*. A system is separated from the remainder of the universe by a *boundary* which may be notional or not, but which by convention delimits a finite volume. Exchanges of work, heat, or matter between the system and the surroundings take place across this boundary.

SURROUNDINGS

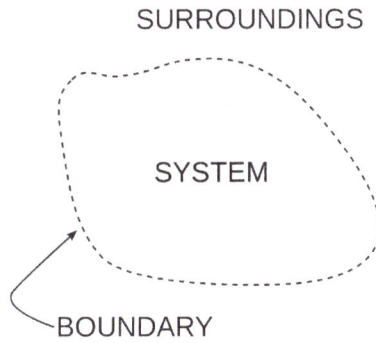

A diagram of a generic thermodynamic system

In practice, the boundary is simply an imaginary dotted line drawn around a volume when there is going to be a change in the internal energy of that volume. Anything that passes across the boundary that affects a change in the internal energy needs to be accounted for in the energy balance equation. The volume can be the region surrounding a single atom resonating energy, such as Max Planck defined in 1900; it can be a body of steam or air in a steam engine, such as Sadi Carnot defined in 1824; it can be the body of a tropical cyclone, such as Kerry Emanuel theorized in 1986 in the field of atmospheric thermodynamics; it could also be just one nuclide (i.e. a system of quarks) as hypothesized in quantum thermodynamics.

Boundaries are of four types: fixed, moveable, real, and imaginary. For example, in an engine, a fixed boundary means the piston is locked at its position; as such, a constant volume process occurs. In that same engine, a moveable boundary allows the piston to move in and out. For closed systems, boundaries are real while for open system boundaries are often imaginary.

Generally, thermodynamics distinguishes three classes of systems, defined in terms of what is allowed to cross their boundaries:

Interactions of thermodynamic systems			
Type of system	**Mass flow**	**Work**	**Heat**
Open	✓	✓	✓
Closed	✗	✓	✓
Thermally isolated	✗	✓	✗
Mechanically isolated	✗	✗	✓
Isolated	✗	✗	✗

As time passes in an isolated system, internal differences in the system tend to even out and pressures and temperatures tend to equalize, as do density differences. A system in which all equalizing processes have gone to completion is considered to be in a state of thermodynamic equilibrium.

In thermodynamic equilibrium, a system's properties are, by definition, unchanging in time. Systems in equilibrium are much simpler and easier to understand than systems which are not in equilibrium. Often, when analysing a thermodynamic process, it can be assumed that each intermediate state in the process is at equilibrium. This will also considerably simplify the situation. Thermodynamic processes which develop so slowly as to allow each intermediate step to be an equilibrium state are said to be reversible processes.

States and Processes

When a system is at equilibrium under a given set of conditions, it is said to be in a definite thermodynamic state. The state of the system can be described by a number of state quantities that do not depend on the process by which the system arrived at its state. They are called intensive variables or extensive variables according to how they change when the size of the system changes. The properties of the system can be described by an equation of state which specifies the relationship between these variables. State may be thought of as the instantaneous quantitative description of a system with a set number of variables held constant.

A thermodynamic process may be defined as the energetic evolution of a thermodynamic system proceeding from an initial state to a final state. It can be described by process quantities. Typically, each thermodynamic process is distinguished from other processes in energetic character according to what parameters, such as temperature, pressure, or volume, etc., are held fixed. Furthermore, it is useful to group these processes into pairs, in which each variable held constant is one member of a conjugate pair.

Several commonly studied thermodynamic processes are:

- Isobaric process: occurs at constant pressure

- Isochoric process: occurs at constant volume (also called isometric/isovolumetric)

- Isothermal process: occurs at a constant temperature

- Adiabatic process: occurs without loss or gain of energy by heat

- Isentropic process: a reversible adiabatic process, occurs at a constant entropy

- Isenthalpic process: occurs at a constant enthalpy

- Steady state process: occurs without a change in the internal energy

Instrumentation

There are two types of thermodynamic instruments, the meter and the reservoir. A thermodynamic meter is any device which measures any parameter of a thermodynamic system. In some cases, the thermodynamic parameter is actually defined in terms of an idealized measuring instrument. For example, the zeroth law states that if two bodies are in thermal equilibrium with a third body, they are also in thermal equilibrium with each other. This principle, as noted by James Maxwell in 1872, asserts that it is possible to measure temperature. An idealized thermometer is a sample of an ideal gas at constant pressure. From the ideal gas law $pV=nRT$, the volume of such a sample can be used as an indicator of temperature; in this manner it defines temperature. Although pressure is defined mechanically, a pressure-measuring device, called a barometer may also be constructed from a sample of an ideal gas held at a constant temperature. A calorimeter is a device which is used to measure and define the internal energy of a system.

A thermodynamic reservoir is a system which is so large that it does not appreciably alter its state parameters when brought into contact with the test system. It is used to impose a particular value of a state parameter upon the system. For example, a pressure reservoir is a system at a particular pressure, which imposes that pressure upon any test system that it is mechanically connected to. The Earth's atmosphere is often used as a pressure reservoir.

Conjugate Variables

The central concept of thermodynamics is that of energy, the ability to do work. By the First Law, the total energy of a system and its surroundings is conserved. Energy may be transferred into a system by heating, compression, or addition of matter, and extracted from a system by cooling, expansion, or extraction of matter. In mechanics, for example, energy transfer equals the product of the force applied to a body and the resulting displacement.

Conjugate variables are pairs of thermodynamic concepts, with the first being akin to a "force" applied to some thermodynamic system, the second being akin to the resulting "displacement," and the product of the two equalling the amount of energy transferred. The common conjugate variables are:

- Pressure-volume (the mechanical parameters);
- Temperature-entropy (thermal parameters);
- Chemical potential-particle number (material parameters).

Potentials

Thermodynamic potentials are different quantitative measures of the stored energy in a system. Potentials are used to measure energy changes in systems as they evolve from an initial state to a final state. The potential used depends on the constraints of the system, such as constant temperature or pressure. For example, the Helmholtz and Gibbs energies are the energies available in a system to do useful work when the temperature and volume or the pressure and temperature are fixed, respectively.

Applied Fields

- Atmospheric thermodynamics

- Biological thermodynamics

- Black hole thermodynamics

- Chemical thermodynamics

- Classical thermodynamics

- Equilibrium thermodynamics

- Industrial ecology (re: Exergy)

- Maximum entropy thermodynamics

- Non-equilibrium thermodynamics

- Philosophy of thermal and statistical physics

- Psychrometrics

- Quantum thermodynamics

- Statistical thermodynamics

- Thermoeconomics

Internal Energy

In thermodynamics, the internal energy of a system is the energy contained within the system, excluding the kinetic energy of motion of the system as a whole and the potential energy of the system as a whole due to external force fields. It keeps account of the gains and losses of energy of the system that are due to changes in its internal state.

The internal energy of a system can be changed by transfers of matter or heat or by doing work. When matter transfer is prevented by impermeable containing walls, the system is said to be closed. Then the first law of thermodynamics states that the increase in internal energy is equal to the total heat added plus the work done on the system by its surroundings. If the containing walls pass neither matter nor energy, the system is said to be isolated and its internal energy cannot change. The first law of thermodynamics may be regarded as establishing the existence of the internal energy.

The internal energy is one of the two cardinal state functions of the state variables of a thermodynamic system.

Introduction

The internal energy of a given state of a system cannot be directly measured. It is determined through some convenient chain of thermodynamic operations and thermodynamic processes by which the given state can be prepared, starting with a reference state which is customarily assigned a reference value for its internal energy. Such a chain, or path, can be theoretically described by certain extensive state variables of the system, namely, its entropy, S, its volume, V, and its mole numbers, $\{N_j\}$. The internal energy, $U(S,V,\{N_j\})$, is a function of those. Sometimes, to that list are appended other extensive state variables, for example electric dipole moment. For practical considerations in thermodynamics and engineering it is rarely necessary or convenient to consider all energies belonging to the total intrinsic energy of a system, such as the energy given by the equivalence of mass. Customarily, thermodynamic descriptions include only items relevant to the processes under study. Thermodynamics is chiefly concerned only with *changes* in the internal energy, not with its absolute value.

The internal energy is a state function of a system, because its value depends only on the current state of the system and not on the path taken or processes undergone to prepare it. It is an extensive quantity. It is the one and only *cardinal* thermodynamic potential. Through it, by use of Legendre transforms, are mathematically constructed the other thermodynamic potentials. These are functions of variable lists in which some extensive variables are replaced by their conjugate intensive variables. Legendre transformation is necessary because mere substitutive replacement of extensive variables by intensive variables does not lead to thermodynamic potentials. Mere substitution leads to a less informative formula, an equation of state.

Though it is a macroscopic quantity, internal energy can be explained in microscopic terms by two theoretical virtual components. One is the microscopic kinetic energy due to the microscopic motion of the system's particles (translations, rotations, vibrations). The other is the potential energy associated with the microscopic forces, including the chemical bonds, between the particles; this is for ordinary physics and chemistry. If thermonuclear reactions are specified as a topic of concern, then the static rest mass energy of the constituents of matter is also counted. There is no simple universal relation between these quantities of microscopic energy and the quantities of energy gained or lost by the system in work, heat, or matter transfer.

The SI unit of energy is the joule (J). Sometimes it is convenient to use a corresponding density called *specific internal energy* which is internal energy per unit of mass (kilogram) of the system in question. The SI unit of specific internal energy is J/kg. If the specific internal energy is expressed relative to units of amount of substance (mol), then it is referred to as *molar internal energy* and the unit is J/mol.

From the standpoint of statistical mechanics, the internal energy is equal to the ensemble average of the sum of the microscopic kinetic and potential energies of the system.

Cardinal Functions

The internal energy, $U(S,V,\{N_j\})$, expresses the thermodynamics of a system in the *energy-language*, or in the *energy representation*. Its arguments are exclusively extensive variables of state. Alongside the internal energy, the other cardinal function of state of a thermodynamic system is its entropy, as a function, $S(U,V,\{N_j\})$, of the same list of extensive variables of state, except that the entropy, S, is replaced in the list by the internal energy, U. It expresses the *entropy representation*.

Each cardinal function is a monotonic function of each of its *natural* or *canonical* variables. Each provides its *characteristic* or *fundamental* equation, for example $U = U(S,V,\{N_j\})$, that by itself contains all thermodynamic information about the system. The fundamental equations for the two cardinal functions can in principle be interconverted by solving, for example, $U = U(S,V,\{N_j\})$ for S, to get $S = S(U,V,\{N_j\})$.

In contrast, Legendre transforms are necessary to derive fundamental equations for other thermodynamic potentials and Massieu functions. The entropy as a function only of extensive state variables is the one and only *cardinal function* of state for the generation of Massieu functions. It is not itself customarily designated a 'Massieu function', though rationally it might be thought of as such, corresponding to the term 'thermodynamic potential', which includes the internal energy.

For real and practical systems, explicit expressions of the fundamental equations are almost always unavailable, but the functional relations exist in principle. Formal, in principle, manipulations of them are valuable for the understanding of thermodynamics.

Description and Definition

The internal energy U of a given state of the system is determined relative to that of a standard state of the system, by adding up the macroscopic transfers of energy that accompany a change of state from the reference state to the given state:

$$\Delta U = \sum_i E_i$$

where ΔU denotes the difference between the internal energy of the given state and that of the reference state, and the E_i are the various energies transferred to the system in the steps from the reference state to the given state. It is the energy needed to create the given state of the system from the reference state.

From a non-relativistic microscopic point of view, it may be divided into microscopic potential energy, $U_{\text{micro pot}}$, and microscopic kinetic energy, $U_{\text{micro kin}}$, components:

$$U = U_{\mathrm{micro\,pot}} + U_{\mathrm{micro\,kin}}$$

The microscopic kinetic energy of a system arises as the sum of the motions of all the system's particles with respect to the center-of-mass frame, whether it be the motion of atoms, molecules, atomic nuclei, electrons, or other particles. The microscopic potential energy algebraic summative components are those of the chemical and nuclear particle bonds, and the physical force fields within the system, such as due to internal induced electric or magnetic dipole moment, as well as the energy of deformation of solids (stress-strain). Usually, the split into microscopic kinetic and potential energies is outside the scope of macroscopic thermodynamics.

Internal energy does not include the energy due to motion or location of a system as a whole. That is to say, it excludes any kinetic or potential energy the body may have because of its motion or location in external gravitational, electrostatic, or electromagnetic fields. It does, however, include the contribution of such a field to the energy due to the coupling of the internal degrees of freedom of the object with the field. In such a case, the field is included in the thermodynamic description of the object in the form of an additional external parameter.

For practical considerations in thermodynamics or engineering, it is rarely necessary, convenient, nor even possible, to consider all energies belonging to the total intrinsic energy of a sample system, such as the energy given by the equivalence of mass. Typically, descriptions only include components relevant to the system under study. Indeed, in most systems under consideration, especially through thermodynamics, it is impossible to calculate the total internal energy. Therefore, a convenient null reference point may be chosen for the internal energy.

The internal energy is an extensive property: it depends on the size of the system, or on the amount of substance it contains.

At any temperature greater than absolute zero, microscopic potential energy and kinetic energy are constantly converted into one another, but the sum remains constant in an isolated system (cf. table). In the classical picture of thermodynamics, kinetic energy vanishes at zero temperature and the internal energy is purely potential energy. However, quantum mechanics has demonstrated that even at zero temperature particles maintain a residual energy of motion, the zero point energy. A system at absolute zero is merely in its quantum-mechanical ground state, the lowest energy state available. At absolute zero a system of given composition has attained its minimum attainable entropy.

The microscopic kinetic energy portion of the internal energy gives rise to the temperature of the system. Statistical mechanics relates the pseudo-random kinetic energy of individual particles to the mean kinetic energy of the entire ensemble of particles comprising a system. Furthermore, it relates the mean microscopic kinetic energy to the macroscopically observed empirical property that is expressed as temperature of the system. This energy is often referred to as the *thermal energy* of a system, relating this energy, like the temperature, to the human experience of hot and cold.

Statistical mechanics considers any system to be statistically distributed across an ensemble of N microstates. Each microstate has an energy E_i and is associated with a probability p_i. The internal energy is the mean value of the system's total energy, i.e., the sum of all microstate energies, each weighted by their probability of occurrence:

$$U = \sum_{i=1}^{N} p_i E_i .$$

This is the statistical expression of the first law of thermodynamics.

Internal Energy Changes

Type of system	Mass flow	Work	Heat
Open	✓	✓	✓
Closed	✗	✓	✓
Thermally isolated	✗	✓	✗
Mechanically isolated	✗	✗	✓
Isolated	✗	✗`	✗

Thermodynamics is chiefly concerned only with the changes, ΔU, in internal energy.

For a closed system, with matter transfer excluded, the changes in internal energy are due to heat transfer Q and due to work. The latter can be split into two kinds, pressure-volume work $W_{\text{pressure-volume}}$, and frictional and other kinds, such as electrical polarization, which do not alter the volume of the system, and are called isochoric,

In an ideal gas all of the extra energy results in a temperature increase, as it is stored solely as microscopic kinetic energy; such heating is said to be *sensible*.

A second mechanism of change of internal energy of a closed system is the doing of work on the system, either in mechanical form by changing pressure or volume, or by other perturbations, such as directing an electric current through the system.

If the system is not closed, the third mechanism that can increase the internal energy is transfer of matter into the system. This increase, ΔU_{matter} cannot be split into heat and work components. If the system is so set up physically that heat and work can be done on it by pathways separate from and independent of matter transfer, then the transfers of energy add to change the internal energy:

$$\Delta U = Q + W_{\text{pressure-volume}} + W_{\text{isochoric}} + \Delta U_{\text{matter}}$$

(separate pathway for matter transfer from heat and work transfer pathways).

If a system undergoes certain phase transformations while being heated, such as melting and vaporization, it may be observed that the temperature of the system does not change until the entire sample has completed the transformation. The energy introduced into the system while the temperature did not change is called a *latent energy*, or latent heat, in contrast to sensible heat, which is associated with temperature change.

Thermodynamic Operation

A thermodynamic operation is an externally imposed manipulation or change of connection or wall between a thermodynamic system and its surroundings. An externally imposed change in the state of the surroundings of a system can usually be regarded as a virtual thermodynamic operation. It may usually be analyzed as due to changes in the walls that separate the systems in the surroundings. It is assumed in thermodynamics that the operation is conducted in ignorance of any pertinent microscopic information.

A thermodynamic operation requires a contribution from an animate agency, or at least an independent external agency, that does not come from the passive or inanimate properties of the systems. Perhaps the first expression of the distinction between a thermodynamic operation and a thermodynamic process is in Kelvin's statement of the second law of thermodynamics: "It is impossible, by means of inanimate material agency, to derive mechanical effect from any portion of matter by cooling it below the temperature of the surrounding objects." A sequence of events that occurred other than "by means of inanimate material agency" would entail an action by an animate agency, or at least an independent external agency. Such an agency could impose some thermodynamic operations. For example, those operations might create a heat pump, which of course would comply with the second law. A Maxwell's demon conducts an extremely idealized and naturally unrealizable kind of thermodynamic operation.

An ordinary language expression for a thermodynamic operation is used by Edward A. Guggenheim: "tampering" with the bodies.

Distinction Between Thermodynamic Operation and Thermodynamic Process

A typical thermodynamic operation is a removal of an initially separating wall, a manipulation that unites two systems into one undivided system. A typical thermodynamic process consists of a redistribution that spreads a conserved quantity between a system and its surroundings across a previously impermeable but newly unchanging semi-permeable wall between them. More generally, a process can be considered as a transfer of some quantity that is defined by an extensive state variable of the system, so that a transfer balance equation can be written. According to Uffink, "... thermodynamic processes only take place after an external intervention on the system (such as: removing a partition, establishing thermal contact with a heat bath, pushing a piston, etc.). They do not correspond to the autonomous behaviour of a free system."

As a matter of history, the distinction, between a thermodynamic operation and a thermodynamic process, is not found in these terms in nineteenth century accounts. For example, Kelvin spoke of a "thermodynamic operation" when he meant what present-day terminology calls a thermodynamic operation followed by a thermodynamic process. Again, Planck usually spoke of a "process" when our present-day terminology would speak of a thermodynamic operation followed by a thermodynamic process.

Planck's "Natural Processes" Contrasted with Actions of Maxwell's Demon

Planck held that all "natural processes" (meaning, in present-day terminology, a thermodynamic operation followed by a thermodynamic process) are irreversible and proceed in the sense of increase of entropy sum. In these terms, it would be by thermodynamic operations that, if he could exist, Maxwell's demon would conduct unnatural affairs, which include transitions in the sense away from thermodynamic equilibrium. They are physically theoretically conceivable up to a point, but are not natural processes in Planck's sense. The reason is that ordinary thermodynamic operations are conducted in total ignorance of the very kinds of microscopic information that is essential to the efforts of Maxwell's demon.

Examples of Thermodynamic Operations

Thermodynamic Cycle

A thermodynamic cycle is constructed as a sequence of stages or steps. Each stage consists of a thermodynamic operation followed by a thermodynamic process. For example, an initial thermodynamic operation of a cycle of a Carnot heat engine could be taken as the setting of the working body, at a known high temperature, into contact with a thermal reservoir at the same temperature (the hot reservoir), through a wall permeable only to heat, while it remains in mechanical contact with the work reservoir. This thermodynamic operation is followed by a thermodynamic process, in which the expansion of the working body is so slow as to be effectively reversible, while internal energy is transferred as heat from the hot reservoir to the working body and as work from the working body to the work reservoir. Theoretically, the process terminates eventually, and this ends the stage. The engine is then subject to another thermodynamic operation, and the cycle proceeds into another stage. The cycle completes when the thermodynamic variables (the thermodynamic state) of the working body return to their initial values.

Virtual Thermodynamic Operations

A refrigeration device passes a working substance through successive stages, overall constituting a cycle. This may be brought about not by moving or changing separating walls around an unmoving body of working substance, but rather by moving a body of working substance to bring about exposure to a cyclic succession of unmoving unchanging walls. The effect is virtually a cycle of thermodynamic operations. The kinetic energy of bulk motion of the working substance is not a significant feature of the device, and the working substance may be practically considered as nearly at rest.

Composition of Systems

For many chains of reasoning in thermodynamics, it is convenient to think of the combination of two systems into one. It is imagined that the two systems, separated from their surroundings, are juxtaposed and (by a shift of viewpoint) regarded as constituting a new, composite system. The composite system is imagined amid its new overall surroundings. This sets up the possibility of interaction between the two subsystems and between the composite system and its overall surroundings, for example by allowing contact through a wall with a particular kind of permeability. This conceptual device was introduced into thermodynamics mainly in the work of Carathéodory, and has been widely used since then.

Additivity of Extensive Variables

If the thermodynamic operation is entire removal of walls, then extensive state variables of the composed system are the respective sums of those of the component systems. This is called the additivity of extensive variables.

Scaling of a System

A thermodynamic system consisting of a single phase, in the absence of external forces, in its own state of internal thermodynamic equilibrium, is homogeneous. This means that the material in any region of the system can be interchanged with the material of any congruent and parallel region of the system, and the effect is to leave the system thermodynamically unchanged. The thermodynamic operation of *scaling* is the creation of a new homogeneous system whose size is a multiple of the old size, and whose intensive variables have the same values. Traditionally the size is stated by the mass of the system, but sometimes it is stated by the entropy, or by the volume. For a given such system Φ, scaled by the real number λ to yield a new one $\lambda\Phi$, a state function, $X(.)$, such that $X(\lambda\Phi) = \lambda X(\Phi)$, is said to be extensive. Such a function as X is called a homogeneous function of degree 1. There are two different concepts mentioned here, sharing the same name: (a) the mathematical concept of degree-1 homogeneity in the scaling function; and (b) the physical concept of the spatial homogeneity of the system. It happens that the two agree here, but that is not because they are tautologous. It is a contingent fact of thermodynamics.

Splitting and Recomposition of Systems

If two systems, S_a and S_b, have identical intensive variables, a thermodynamic operation of wall removal can compose them into a single system, S, with the same intensive variables. If, for example, their internal energies are in the ratio $\lambda:(1-\lambda)$, then the composed system, S, has internal energy in the ratio of $1:\lambda$ to that of the system S_a. By the inverse thermodynamic operation, the system S can be split into two subsystems in the obvious way. As usual, these thermodynamic operations are conducted in total ignorance of the microscopic states of the systems. More particularly, it is characteristic of macroscopic thermodynamics that the probability vanishes, that the splitting operation occurs at an instant when system S is in the kind of extreme transient microscopic state envisaged by the Poincaré recurrence argument. Such splitting and recomposition is in accord with the above defined additivity of extensive variables.

Statements of Laws

Thermodynamic operations appear in the statements of the laws of thermodynamics. For the zeroth law, one considers operations of thermally connecting and disconnecting systems. For the second law, some statements contemplate an operation of connecting two initially unconnected systems. For the third law, one statement is that no finite sequence of thermodynamic operations can bring a system to absolute zero temperature.

Equilibrium Thermodynamics

Equilibrium Thermodynamics is the systematic study of transformations of matter and energy in systems in terms of a concept called thermodynamic equilibrium. The word equilibrium implies a state of balance. Equilibrium thermodynamics, in origins, derives from analysis of the Carnot cycle. Here, typically a system, as cylinder of gas, initially in its own state of internal thermodynamic equilibrium, is set *out of balance* via heat input from a combustion reaction. Then, through a series of steps, as the system settles into its final equilibrium state, work is extracted.

In an equilibrium state the potentials, or driving forces, within the system, are in exact balance. A central aim in equilibrium thermodynamics is: given a system in a well-defined initial state of thermodynamic equilibrium, subject to accurately specified constraints, to calculate, when the constraints are changed by an externally imposed intervention, what the state of the system will be once it has reached a new equilibrium. An equilibrium state is mathematically ascertained by seeking the extrema of a thermodynamic potential function, whose nature depends on the constraints imposed on the system. For example, a chemical reaction at constant temperature and pressure will reach equilibrium at a minimum of its components' Gibbs free energy and a maximum of their entropy.

Equilibrium thermodynamics differs from non-equilibrium thermodynamics, in that, with the latter, the state of the system under investigation will typically not be uniform but will vary locally in those as energy, entropy, and temperature distributions as gradients are imposed by dissipative thermodynamic fluxes. In equilibrium thermodynamics, by contrast, the state of the system will be considered uniform throughout, defined macroscopically by such quantities as temperature, pressure, or volume. Systems are studied in terms of change from one equilibrium state to another; such a change is called a thermodynamic process.

Ruppeiner geometry is a type of information geometry used to study thermodynamics. It claims that thermodynamic systems can be represented by Riemannian geometry, and that statistical properties can be derived from the model. This geometrical model is based on the idea that there exist equilibrium states which can be represented by points on two-dimensional surface and the distance between these equilibrium states is related to the fluctuation between them.

Non-equilibrium Thermodynamics

Non-equilibrium thermodynamics is a branch of thermodynamics that deals with physical systems that are not in thermodynamic equilibrium but can be adequately described in terms of variables (non-equilibrium state variables) that represent an extrapolation of the variables used to specify the system in thermodynamic equilibrium. Non-equilibrium thermodynamics is concerned with transport processes and with the rates of chemical reactions. It relies on what may be thought of as more or less nearness to thermodynamic equilibrium. Non-equilibrium thermodynamics is a work in progress, not an established edifice. This article will try to sketch some approaches to it and some concepts important for it.

Almost all systems found in nature are not in thermodynamic equilibrium; for they are changing or can be triggered to change over time, and are continuously and discontinuously subject to flux of matter and energy to and from other systems and to chemical reactions. Some systems and processes are, however, in a useful sense, near enough to thermodynamic equilibrium to allow description with useful accuracy by currently known non-equilibrium thermodynamics. Nevertheless, many natural systems and processes will always remain far beyond the scope of non-equilibrium thermodynamic methods. This is because of the very small size of atoms, as compared with macroscopic systems.

The thermodynamic study of non-equilibrium systems requires more general concepts than are dealt with by equilibrium thermodynamics. One fundamental difference between equilibrium thermodynamics and non-equilibrium thermodynamics lies in the behaviour of inhomogeneous systems, which require for their study knowledge of rates of reaction which are not considered in equilibrium thermodynamics of homogeneous systems. This is discussed below. Another fundamental and very important difference is the difficulty or impossibility in defining entropy at an instant of time in macroscopic terms for systems not in thermodynamic equilibrium.

Scope of Non-equilibrium Thermodynamics

Difference Between Equilibrium and Non-equilibrium Thermodynamics

A profound difference separates equilibrium from non-equilibrium thermodynamics. Equilibrium thermodynamics ignores the time-courses of physical processes. In contrast, non-equilibrium thermodynamics attempts to describe their time-courses in continuous detail.

Equilibrium thermodynamics restricts its considerations to processes that have initial and final states of thermodynamic equilibrium; the time-courses of processes are deliberately ignored. Consequently, equilibrium thermodynamics allows processes that pass through states far from thermodynamic equilibrium, that cannot be described even by the variables admitted for non-equilibrium thermodynamics, such as time rates of change of temperature and pressure. For example, in equilibrium thermodynamics, a process is allowed to include even a violent explosion that cannot be described by non-equilibrium thermodynamics. Equilibrium thermodynamics does, however, for theoretical development, use the idealized concept of the "quasi-static process". A quasi-static process is a conceptual (timeless and physically impossible) smooth mathematical passage along a continuous path of states of thermodynamic equilibrium. It is an exercise in differential geometry rather than a process that could occur in actuality.

Non-equilibrium thermodynamics, on the other hand, attempting to describe continuous time-courses, need its state variables to have a very close connection with those of equilibrium thermodynamics. This profoundly restricts the scope of non-equilibrium thermodynamics, and places heavy demands on its conceptual framework.

Non-equilibrium State Variables

The suitable relationship that defines non-equilibrium thermodynamic state variables is as follows. On occasions when the system happens to be in states that are sufficiently close to thermodynamic equilibrium, non-equilibrium state variables are such that they can be measured locally with sufficient accuracy by the same techniques as are used to measure thermodynamic state variables, or by corresponding time and space derivatives, including fluxes of matter and energy. In general, non-equilibrium thermodynamic systems are spatially and temporally non-uniform, but their non-uniformity still has a sufficient degree of smoothness to support the existence of suitable time and space derivatives of non-equilibrium state variables. Because of the spatial non-uniformity, non-equilibrium state variables that correspond to extensive thermodynamic state variables have to be defined as spatial densities of the corresponding extensive equilibrium state variables. On occasions when the system is sufficiently close to thermodynamic equilibrium, intensive non-equilibrium state variables, for example temperature and pressure, correspond closely with equilibrium state variables. It is necessary that measuring probes be small enough, and rapidly enough responding, to capture relevant non-uniformity. Further, the non-equilibrium state variables are required to be mathematically functionally related to one another in ways that suitably resemble corresponding relations between equilibrium thermodynamic state variables. In reality, these requirements are very demanding, and it may be difficult or practically, or even theoretically, impossible to satisfy them. This is part of why non-equilibrium thermodynamics is a work in progress.

Overview

Non-equilibrium thermodynamics is a work in progress, not an established edifice. This article will try to sketch some approaches to it and some concepts important for it.

Some concepts of particular importance for non-equilibrium thermodynamics include time rate of dissipation of energy (Rayleigh 1873, Onsager 1931, also), time rate of entropy production (Onsager 1931), thermodynamic fields, dissipative structure, and non-linear dynamical structure.

One problem of interest is the thermodynamic study of non-equilibrium steady states, in which entropy production and some flows are non-zero, but there is no time variation of physical variables.

One initial approach to non-equilibrium thermodynamics is sometimes called 'classical irreversible thermodynamics'. There are other approaches to non-equilibrium thermodynamics, for example extended irreversible thermodynamics, and generalized thermodynamics, but they are hardly touched on in the present article.

Quasi-radiationless Non-equilibrium Thermodynamics of Matter in Laboratory Conditions

According to Wildt, current versions of non-equilibrium thermodynamics ignore radiant heat; they can do so because they refer to laboratory quantities of matter under laboratory conditions with temperatures well below those of stars. At laboratory temperatures, in laboratory quantities of matter, thermal radiation is weak and can be practically nearly ignored. But, for example, atmospheric physics is concerned with large amounts of matter, occupying cubic kilometers, that, taken as a whole, are not within the range of laboratory quantities; then thermal radiation cannot be ignored.

Local Equilibrium Thermodynamics

The terms 'classical irreversible thermodynamics' and 'local equilibrium thermodynamics' are sometimes used to refer to a version of non-equilibrium thermodynamics that demands certain simplifying assumptions, as follows. The assumptions have the effect of making each very small volume element of the system effectively homogeneous, or well-mixed, or without an effective spatial structure, and without kinetic energy of bulk flow or of diffusive flux. Even within the thought-frame of classical irreversible thermodynamics, care is needed in choosing the independent variables for systems. In some writings, it is assumed that the intensive variables of equilibrium thermodynamics are sufficient as the independent variables for the task; in particular, local flow intensive variables are not admitted as independent variables; local flows are considered as dependent on quasi-static local intensive variables.

Also it is assumed that the local entropy density is the same function of the other local intensive variables as in equilibrium; this is called the local thermodynamic equilibrium assumption. Radiation is ignored because it is transfer of energy between regions, which can be remote from one another. In the classical irreversible thermodynamic approach, there is allowed very small spatial variation, from very small volume element to adjacent very small volume element, but it is assumed that the global entropy of the system can be found by simple spatial integration of the local entropy density; this means that spatial structure cannot contribute as it properly should to the global entropy assessment for the system. This approach assumes spatial and temporal continuity and even differentiability of locally defined intensive variables such as temperature and internal energy density. All of these are very stringent demands. Consequently, this approach can deal with only a very limited range of phenomena. This approach is nevertheless valuable because it can deal well with some macroscopically observable phenomena.

In other writings, local flow variables are considered; these might be considered as classical by analogy with the time-invariant long-term time-averages of flows produced by endlessly repeated cyclic processes; examples with flows are in the thermoelectric phenomena known as the Seebeck and the Peltier effects, considered by Kelvin in the nineteenth century and by Onsager in the twentieth. These effects occur at metal junctions, which were originally effectively treated as two-dimensional surfaces, with no spatial volume, and no spatial variation.

Local Equilibrium Thermodynamics with Materials with "Memory"

A further extension of local equilibrium thermodynamics is to allow that materials may have "memory", so that their constitutive equations depend not only on present values but also on past values of local equilibrium variables. Thus time comes into the picture more deeply than for time-dependent local equilibrium thermodynamics with memoryless materials, but fluxes are not independent variables of state.

Extended Irreversible Thermodynamics

Extended irreversible thermodynamics is a branch of non-equilibrium thermodynamics that goes outside the restriction to the local equilibrium hypothesis. The space of state variables is enlarged by including the fluxes of mass, momentum and energy and eventually higher order fluxes. The formalism is well-suited for describing high-frequency processes and small-length scales materials.

Basic Concepts

There are many examples of stationary non-equilibrium systems, some very simple, like a system confined between two thermostats at different temperatures or the ordinary Couette flow, a fluid enclosed between two flat walls moving in opposite directions and defining non-equilibrium conditions at the walls. Laser action is also a non-equilibrium process, but it depends on departure from local thermodynamic equilibrium and is thus beyond the scope of classical irreversible thermodynamics; here a strong temperature difference is maintained between two molecular degrees of freedom (with molecular laser, vibrational and rotational molecular motion), the requirement for two component 'temperatures' in the one small region of space, precluding local thermodynamic equilibrium, which demands that only one temperature be needed. Damping of acoustic perturbations or shock waves are non-stationary non-equilibrium processes. Driven complex fluids, turbulent systems and glasses are other examples of non-equilibrium systems.

The mechanics of macroscopic systems depends on a number of extensive quantities. It should be stressed that all systems are permanently interacting with their surroundings, thereby causing unavoidable fluctuations of extensive quantities. Equilibrium conditions of thermodynamic systems are related to the maximum property of the entropy. If the only extensive quantity that is allowed to fluctuate is the internal energy, all the other ones being kept strictly constant, the temperature of the system is measurable and meaningful. The system's properties are then most conveniently described using the thermodynamic potential Helmholtz free energy ($A = U - TS$), a Legendre transformation of the energy. If, next to fluctuations of the energy, the macroscopic dimensions (volume) of the system are left fluctuating, we use the Gibbs free energy ($G = U + PV - TS$), where the system's properties are determined both by the temperature and by the pressure.

Non-equilibrium systems are much more complex and they may undergo fluctuations of more extensive quantities. The boundary conditions impose on them particular intensive variables, like temperature gradients or distorted collective motions (shear motions, vortices, etc.), often called thermodynamic forces. If free energies are very useful in equilibrium thermodynamics, it must be stressed that there is no general law defining stationary non-equilibrium properties of the energy as is the second law of thermodynamics for the entropy in equilibrium thermodynamics. That is why in such cases a more generalized Legendre transformation should be considered. This is the extended Massieu potential. By definition, the entropy (S) is a function of the collection of extensive quantities E_i. Each extensive quantity has a conjugate intensive variable I_i (a restricted definition of intensive variable is used here by comparison to the definition given in this link) so that:

$$I_i = \partial S / \partial E_i.$$

We then define the extended Massieu function as follows:

$$k_b M = S - \sum_i (I_i E_i),$$

where k_b is Boltzmann's constant, whence

$$k_b dM = \sum_i (E_i dI_i).$$

The independent variables are the intensities.

Intensities are global values, valid for the system as a whole. When boundaries impose to the system different local conditions, (e.g. temperature differences), there are intensive variables representing the average value and others representing gradients or higher moments. The latter are the thermodynamic forces driving fluxes of extensive properties through the system.

It may be shown that the Legendre transformation changes the maximum condition of the entropy (valid at equilibrium) in a minimum condition of the extended Massieu function for stationary states, no matter whether at equilibrium or not.

Stationary States, Fluctuations, and Stability

In thermodynamics one is often interested in a stationary state of a process, allowing that the stationary state include the occurrence of unpredictable and experimentally unreproducible fluctuations in the state of the system. The fluctuations are due to the system's internal sub-processes and to exchange of matter or energy with the system's surroundings that create the constraints that define the process.

If the stationary state of the process is stable, then the unreproducible fluctuations involve local transient decreases of entropy. The reproducible response of the system is then to increase the entropy back to its maximum by irreversible processes: the fluctuation cannot be reproduced with a significant level of probability. Fluctuations about stable stationary states are extremely small except near critical points (Kondepudi and Prigogine 1998, page 323). The stable stationary state has a local maximum of entropy and is locally the most reproducible state of the system. There are theorems about the irreversible dissipation of fluctuations. Here 'local' means local with respect to the abstract space of thermodynamic coordinates of state of the system.

If the stationary state is unstable, then any fluctuation will almost surely trigger the virtually explosive departure of the system from the unstable stationary state. This can be accompanied by increased export of entropy.

Local Thermodynamic Equilibrium

The scope of present-day non-equilibrium thermodynamics does not cover all physical processes. A condition for the validity of many studies in non-equilibrium thermodynamics of matter is that they deal with what is known as *local thermodynamic equilibrium*.

Local Thermodynamic Equilibrium of Ponderable Matter

Local thermodynamic equilibrium of matter means that conceptually, for study and analysis, the system can be spatially and temporally divided into 'cells' or 'micro-phases' of small (infinitesimal) size, in which classical thermodynamical equilibrium conditions for matter are fulfilled to good approximation. These conditions are unfulfilled, for example, in very rarefied gases, in which molecular collisions are infrequent; and in the boundary layers of a star, where radiation is passing energy to space; and for interacting fermions at very low temperature, where dissipative processes become ineffective. When these 'cells' are defined, one admits that matter and energy may pass freely between contiguous 'cells', slowly enough to leave the 'cells' in their respective individual local thermodynamic equilibria with respect to intensive variables.

One can think here of two 'relaxation times' separated by order of magnitude. The longer relaxation time is of the order of magnitude of times taken for the macroscopic dynamical structure of the system to change. The shorter is of the order of magnitude of times taken for a single 'cell' to reach local thermodynamic equilibrium. If these two relaxation times are not well separated, then the classical non-equilibrium thermodynamical concept of local thermodynamic equilibrium loses its meaning and other approaches have to be proposed, see for instance Extended irreversible thermodynamics. For example, in the atmosphere, the speed of sound is much greater than the wind speed; this favours the idea of local thermodynamic equilibrium of matter for atmospheric heat transfer studies at altitudes below about 60 km where sound propagates, but not above 100 km, where, because of the paucity of intermolecular collisions, sound does not propagate.

Milne's 1928 Definition of Local Thermodynamic Equilibrium in Terms of Radiative Equilibrium

Milne (1928), thinking about stars, gave a definition of 'local thermodynamic equilibrium' in terms of the thermal radiation of the matter in each small local 'cell'. He defined 'local thermodynamic equilibrium' in a 'cell' by requiring that it macroscopically absorb and spontaneously emit radiation as if it were in radiative equilibrium in a cavity at the temperature of the matter of the 'cell'. Then it strictly obeys Kirchhoff's law of equality of radiative emissivity and absorptivity, with a black body source function. The key to local thermodynamic equilibrium here is that the rate of collisions of ponderable matter particles such as molecules should far exceed the rates of creation and annihilation of photons.

Entropy in Evolving Systems

It is pointed out by W.T. Grandy Jr that entropy, though it may be defined for a non-equilibrium system, is when strictly considered, only a macroscopic quantity that refers to the whole system, and is not a dynamical variable and in general does not act as a local potential that describes local physical forces. Under special circumstances, however, one can metaphorically think as if the thermal variables behaved like local physical forces. The approximation that constitutes classical irreversible thermodynamics is built on this metaphoric thinking.

This point of view shares many points in common with the concept and the use of entropy in continuum thermomechanics, which evolved completely independently of statistical mechanics and maximum-entropy principles.

The Onsager Relations

Following Section III of Rayleigh (1873), Onsager (1931, I) showed that in the regime where both the flows (J_i) are small and the thermodynamic forces (F_i) vary slowly, the rate of creation of entropy (σ) is linearly related to the flows:

$$\sigma = \sum_i J_i \frac{\partial F_i}{\partial x_i}$$

and the flows are related to the gradient of the forces, parametrized by a matrix of coefficients conventionally denoted L:

$$J_i = \sum_j L_{ij} \frac{\partial F_i}{\partial x_j}$$

from which it follows that:

$$\sigma = \sum_{i,j} L_{ij} \frac{\partial F_i}{\partial x_i} \frac{\partial F_j}{\partial x_j}$$

The second law of thermodynamics requires that the matrix L be positive definite. Statistical mechanics considerations involving microscopic reversibility of dynamics imply that the matrix L is symmetric. This fact is called the *Onsager reciprocal relations*.

Speculated Extremal Principles for Non-equilibrium Processes

Until recently, prospects for useful extremal principles in this area have seemed clouded. C. Nicolis (1999) concludes that one model of atmospheric dynamics has an attractor which is not a regime of maximum or minimum dissipation; she says this seems to rule out the existence of a global organizing principle, and comments that this is to some extent disappointing; she also points to the difficulty of finding a thermodynamically consistent form of entropy production. Another top expert offers an extensive discussion of the possibilities for principles of extrema of entropy production and of dissipation of energy: Chapter 12 of Grandy (2008) is very cautious, and finds difficulty in defining the 'rate of internal entropy production' in many cases, and finds that sometimes for the prediction of the course of a process, an extremum of the quantity called the rate of dissipation of energy may be more useful than that of the rate of entropy production; this quantity appeared in Onsager's 1931 origination of this subject. Other writers have also felt that prospects for general global extremal principles are clouded. Such writers include Glansdorff and Prigogine (1971), Lebon, Jou and Casas-Vásquez (2008), and Šilhavý (1997).

A recent proposal may perhaps by-pass those clouded prospects.

Applications of Non-equilibrium Thermodynamics

Non-equilibrium thermodynamics has been successfully applied to describe biological processes such as protein folding/unfolding and transport through membranes.

Also, ideas from non-equilibrium thermodynamics and the informatic theory of entropy have been adapted to describe general economic systems.

Chemical Thermodynamics

Chemical thermodynamics is the study of the interrelation of heat and work with chemical reactions or with physical changes of state within the confines of the laws of thermodynamics. Chemical thermodynamics involves not only laboratory measurements of various thermodynamic properties, but also the application of mathematical methods to the study of chemical questions and the *spontaneity* of processes.

The structure of chemical thermodynamics is based on the first two laws of thermodynamics. Starting from the first and second laws of thermodynamics, four equations called the "fundamental equations of Gibbs" can be derived. From these four, a multitude of equations, relating the thermodynamic properties of the thermodynamic system can be derived using relatively simple mathematics. This outlines the mathematical framework of chemical thermodynamics.

History

In 1865, the German physicist Rudolf Clausius, in his *Mechanical Theory of Heat*, suggested that the principles of thermochemistry, e.g. the heat evolved in combustion reactions, could be applied to the principles of thermodynamics. Building on the work of Clausius, between the years 1873-76 the American mathematical physicist Willard Gibbs published a series of three papers, the most famous one being the paper *On the Equilibrium of Heterogeneous Substances*. In these papers, Gibbs showed how the first two laws of thermodynamics could be measured graphically and mathematically to determine both the thermodynamic equilibrium of chemical reactions as well as their tendencies to occur or proceed. Gibbs' collection of papers provided the first unified body of thermodynamic theorems from the principles developed by others, such as Clausius and Sadi Carnot.

J. Willard Gibbs - founder of chemical thermodynamics

During the early 20th century, two major publications successfully applied the principles developed by Gibbs to chemical processes, and thus established the foundation of the science of chemical thermodynamics. The first was the 1923 textbook *Thermodynamics and the Free Energy of Chemical Substances* by Gilbert N. Lewis and Merle Randall. This book was responsible for supplanting the chemical affinity with the term free energy in the English-speaking world. The second was the 1933 book *Modern Thermodynamics by the methods of Willard Gibbs* written by E. A. Guggenheim. In this manner, Lewis, Randall, and Guggenheim are considered as the founders of modern chemical thermodynamics because of the major contribution of these two books in unifying the application of thermodynamics to chemistry.

Overview

The primary objective of chemical thermodynamics is the establishment of a criterion for the determination of the feasibility or spontaneity of a given transformation. In this manner, chemical thermodynamics is typically used to predict the energy exchanges that occur in the following processes:

1. Chemical reactions

2. Phase changes

3. The formation of solutions

The following state functions are of primary concern in chemical thermodynamics:

- Internal energy (U)

- Enthalpy (H)

- Entropy (S)

- Gibbs free energy (G)

Most identities in chemical thermodynamics arise from application of the first and second laws of thermodynamics, particularly the law of conservation of energy, to these state functions.

The 3 laws of thermodynamics:

1. The energy of the universe is constant.

2. In any spontaneous process, there is always an increase in entropy of the universe

3. The entropy of a perfect crystal(well ordered) at 0 Kelvin is zero

Chemical Energy

Chemical energy is the potential of a chemical substance to undergo a transformation through a chemical reaction or to transform other chemical substances. Breaking or making of chemical bonds involves energy or heat, which may be either absorbed or evolved from a chemical system.

Energy that can be released (or absorbed) because of a reaction between a set of chemical substances is equal to the difference between the energy content of the products and the reactants. This change in energy is called the change in internal energy of a chemical reaction. Where $\Delta U^{\circ}_{f\,\text{reactants}}$ is the internal energy of formation of the reactant molecules that can be calculated from the bond energies of the various chemical bonds of the molecules under consideration and $\Delta U^{\circ}_{f\,\text{products}}$ is the internal energy of formation of the product molecules. The change in internal energy is a process which is equal to the heat change if it is measured under conditions of constant volume (at STP condition), as in a closed rigid container such as a bomb calorimeter. However, under conditions of constant pressure, as in reactions in vessels open to the atmosphere, the measured heat change is not always equal to the internal energy change, because pressure-volume work also releases or absorbs energy. (The heat change at constant pressure is called the enthalpy change; in this case the enthalpy of formation).

Another useful term is the heat of combustion, which is the energy released due to a combustion reaction and often applied in the study of fuels. Food is similar to hydrocarbon fuel and carbohydrate fuels, and when it is oxidized, its caloric content is similar.

In chemical thermodynamics the term used for the chemical potential energy is chemical potential, and for chemical transformation an equation most often used is the Gibbs-Duhem equation.

Chemical Reactions

In most cases of interest in chemical thermodynamics there are internal degrees of freedom and processes, such as chemical reactions and phase transitions, which always create entropy unless they are at equilibrium, or are maintained at a "running equilibrium" through "quasi-static" changes by being coupled to constraining devices, such as pistons or electrodes, to deliver and receive external work. Even for homogeneous "bulk" materials, the free energy functions depend on the composition, as do all the extensive thermodynamic potentials, including the internal energy. If the quantities $\{N_i\}$, the number of chemical species, are omitted from the formulae, it is impossible to describe compositional changes.

Gibbs Function or Gibbs Energy

For a "bulk" (unstructured) system they are the last remaining extensive variables. For an unstructured, homogeneous "bulk" system, there are still various *extensive* compositional variables $\{N_i\}$ that G depends on, which specify the composition, the amounts of each chemical substance, expressed as the numbers of molecules present or (dividing by Avogadro's number = 6.023×10^{23}), the numbers of moles

$$G = G(T, P, \{N_i\}).$$

For the case where only PV work is possible

$$dG = -SdT + VdP + \sum_i \mu_i dN_i$$

in which μ_i is the chemical potential for the i-th component in the system

$$\mu_i = \left(\frac{\partial G}{\partial N_i} \right)_{T,P,N_{j\neq i},etc.}.$$

The expression for dG is especially useful at constant T and P, conditions which are easy to achieve experimentally and which approximates the condition in living creatures

$$(dG)_{T,P} = \sum_i \mu_i dN_i.$$

Chemical Affinity

While this formulation is mathematically defensible, it is not particularly transparent since one does not simply add or remove molecules from a system. There is always a *process* involved in changing the composition; e.g., a chemical reaction (or many), or movement of molecules from one phase (liquid) to another (gas or solid). We should find a notation which does not seem to imply that the amounts of the components (N_i) can be changed independently. All real processes obey conservation of mass, and in addition, conservation of the numbers of atoms of each kind. Whatever molecules are transferred to or from should be considered part of the "system".

Consequently, we introduce an explicit variable to represent the degree of advancement of a process, a progress variable ξ for the extent of reaction and to the use of the partial derivative $\partial G/\partial \xi$ (in place of the widely used "ΔG", since the quantity at issue is not a finite change). The result is an understandable expression for the dependence of dG on chemical reactions (or other processes). If there is just one reaction

$$(dG)_{T,P} = \left(\frac{\partial G}{\partial \xi} \right)_{T\,P} d\xi.$$

If we introduce the *stoichiometric coefficient* for the *i-th* component in the reaction

$$\nu_i = \partial N_i / \partial \xi$$

which tells how many molecules of *i* are produced or consumed, we obtain an algebraic expression for the partial derivative

$$\left(\frac{\partial G}{\partial \xi} \right)_{T,P} = \sum_i \mu_i \nu_i = -\mathbb{A}$$

where we introduce a concise and historical name for this quantity, the "affinity", symbolized by A, as introduced by Théophile de Donder in 1923. The minus sign comes from the fact the affinity was defined to represent the rule that spontaneous changes will ensue only when the change in the Gibbs free energy of the process is negative, meaning that the chemical species have a positive affinity for each other. The differential for G takes on a simple form which displays its dependence on compositional change

$$(dG)_{T,P} = -\mathbb{A}d\xi.$$

If there are a number of chemical reactions going on simultaneously, as is usually the case

$$(dG)_{T,P} = -\sum_k \mathbb{A}_k d\xi_k.$$

a set of reaction coordinates $\{ \xi_j \}$, avoiding the notion that the amounts of the components (N_i) can be changed independently. The expressions above are equal to zero at thermodynamic equilibrium, while in the general case for real systems, they are negative because all chemical reactions proceeding at a finite rate produce entropy. This can be made even more explicit by introducing the reaction *rates* $d\xi_j/dt$. For each and every physically independent process.

$$\mathbb{A}\,\dot{\xi} \leq 0.$$

This is a remarkable result since the chemical potentials are intensive system variables, depending only on the local molecular milieu. They cannot "know" whether the temperature and pressure (or any other system variables) are going to be held constant over time. It is a purely local criterion and must hold regardless of any such constraints. Of course, it could have been obtained by taking partial derivatives of any of the other fundamental state functions, but nonetheless is a general criterion for ($-T$ times) the entropy production from that spontaneous process; or at least any part of it that is not captured as external work.

We now relax the requirement of a homogeneous "bulk" system by letting the chemical potentials and the affinity apply to any locality in which a chemical reaction (or any other process) is occurring. By accounting for the entropy production due to irreversible processes, the inequality for dG is now replaced by an equality

$$dG = -SdT + VdP - \sum_k \mathbb{A}_k d\xi_k + W'$$

or

$$dG_{T,P} = -\sum_k \mathbb{A}_k d\xi_k + W'.$$

Any decrease in the Gibbs function of a system is the upper limit for any isothermal, isobaric work that can be captured in the surroundings, or it may simply be dissipated, appearing as T times a corresponding increase in the entropy of the system and/or its surrounding. Or it may go partly toward doing external work and partly toward creating entropy. The important point is that the *extent of reaction* for a chemical reaction may be coupled to the displacement of some external mechanical or electrical quantity in such a way that one can advance only if the other one also does. The coupling may occasionally be *rigid*, but it is often flexible and variable.

Solutions

In solution chemistry and biochemistry, the Gibbs free energy decrease ($\partial G/\partial \xi$, in molar units, denoted cryptically by ΔG) is commonly used as a surrogate for ($-T$ times) the entropy produced by spontaneous chemical reactions in situations where there is no work being done; or at least no "useful" work; i.e., other than perhaps some $\pm PdV$. The assertion that all *spontaneous reactions have a negative ΔG* is merely a restatement of the fundamental thermodynamic relation, giving it the physical dimensions of energy and somewhat obscuring its significance in terms of entropy. When there is no useful work being done, it would be less misleading to use the Legendre transforms of the entropy appropriate for constant T, or for constant T and P, the Massieu functions $-F/T$ and $-G/T$ respectively.

Non Equilibrium

Generally the systems treated with the conventional chemical thermodynamics are either at equilibrium or near equilibrium. Ilya Prigogine developed the thermodynamic treatment of open systems that are far from equilibrium. In doing so he has discovered phenomena and structures of completely new and completely unexpected types. His generalized, nonlinear and irreversible thermodynamics has found surprising applications in a wide variety of fields.

The non equilibrium thermodynamics has been applied for explaining how ordered structures e.g. the biological systems, can develop from disorder. Even if Onsager's relations are utilized, the classical principles of equilibrium in thermodynamics still show that linear systems close to equilibrium always develop into states of disorder which are stable to perturbations and cannot explain the occurrence of ordered structures.

Prigogine called these systems dissipative systems, because they are formed and maintained by the dissipative processes which take place because of the exchange of energy between the system and its environment and because they disappear if that exchange ceases. They may be said to live in symbiosis with their environment.

The method which Prigogine used to study the stability of the dissipative structures to perturbations is of very great general interest. It makes it possible to study the most varied problems, such as city traffic problems, the stability of insect communities, the development of ordered biological structures and the growth of cancer cells to mention but a few examples.

System Constraints

In this regard, it is crucial to understand the role of walls and other *constraints*, and the distinction between *independent* processes and *coupling*. Contrary to the clear implications of many reference sources, the previous analysis is not restricted to homogeneous, isotropic bulk systems which can deliver only PdV work to the outside world, but applies even to the most structured systems. There are complex systems with many chemical "reactions" going on at the same time, some of which are really only parts of the same, overall process. An *independent* process is one that *could* proceed even if all others were unaccountably stopped in their tracks. Understanding this is perhaps a "thought experiment" in chemical kinetics, but actual examples exist.

A gas reaction which results in an increase in the number of molecules will lead to an increase in volume at constant external pressure. If it occurs inside a cylinder closed with a piston, the equilibrated reaction can proceed only by doing work against an external force on the piston. The extent variable for the reaction can increase only if the piston moves, and conversely, if the piston is pushed inward, the reaction is driven backwards.

Similarly, a redox reaction might occur in an electrochemical cell with the passage of current in wires connecting the electrodes. The half-cell reactions at the electrodes are constrained if no current is allowed to flow. The current might be dissipated as joule heating, or it might in turn run an electrical device like a motor doing mechanical work. An automobile lead-acid battery can be recharged, driving the chemical reaction backwards. In this case as well, the reaction is not an independent process. Some, perhaps most, of the Gibbs free energy of reaction may be delivered as external work.

The hydrolysis of ATP to ADP and phosphate can drive the force times distance work delivered by living muscles, and synthesis of ATP is in turn driven by a redox chain in mitochondria and chloroplasts, which involves the transport of ions across the membranes of these cellular organelles. The coupling of processes here, and in the previous examples, is often not complete. Gas can leak slowly past a piston, just as it can slowly leak out of a rubber balloon. Some reaction may occur in a battery even if no external current is flowing. There is usually a coupling coefficient, which may depend on relative rates, which determines what percentage of the driving free energy is turned into external work, or captured as "chemical work"; a misnomer for the free energy of another chemical process.

State Function

In thermodynamics, a state function or function of state is a function defined for a system relating several *state variables* or *state quantities* that depends only on the current equilibrium state of the system. State functions do not depend on the path by which the system arrived at its present state. A state function describes the equilibrium state of a system.

For example, internal energy, enthalpy, and entropy are state quantities because they describe quantitatively an equilibrium state of a thermodynamic system, irrespective of how the system arrived in that state. In contrast, mechanical work and heat are process quantities or path functions, because their values depend on the specific *transition* (or path) between two equilibrium states. The mode of description breaks down for quantities exhibiting hysteresis effects.

History

It is likely that the term "functions of state" was used in a loose sense during the 1850s and 60s by those such as Rudolf Clausius, William Rankine, Peter Tait, William Thomson, and it is clear that by the 1870s the term had acquired a use of its own. In 1873, for example, Willard Gibbs, in his paper "Graphical Methods in the Thermodynamics of Fluids", states: "The quantities V, B, T, U, and S are determined when the state of the body is given, and it may be permitted to call them *functions of the state of the body*.

Overview

A thermodynamic system is described by a number of thermodynamic parameters (e.g. temperature, volume, pressure) which are not necessarily independent. The number of parameters needed to describe the system is the dimension of the state space of the system (D). For example, a monatomic gas having a fixed number of particles is a simple case of a two-dimensional system ($D = 2$). In this example, any system is uniquely specified by two parameters, such as pressure and volume, or perhaps pressure and temperature. These choices are equivalent. They are simply different coordinate systems in the two-dimensional thermodynamic state space. Given pressure and temperature, the volume is calculable from them. Likewise, given pressure and volume, the temperature is calculable from them. An analogous statement holds for higher-dimensional spaces, as described by the state postulate.

Quite generally, a state function is on the form

$$F(P,V,T,\ldots) = 0,$$

where P denotes pressure, T denotes temperature, V denotes volume, and the ellipsis denotes other possible state variables like particle number N and entropy S. If the state space is two-dimensional as in the above example, one may visualize the state space as a three-dimensional graph (a surface in three-dimensional space). The labels of the axes are not generally unique, since there are more state variables than three in this case, and any two independent variables suffice to define the state.

When a system changes state continuously, it traces out a "path" in the state space. The path can be specified by noting the values of the state parameters as the system traces out the path, perhaps as a function of time, or some other external variable. For example, we might have the pressure $P(t)$ and the volume $V(t)$ as functions of time from time t_0 to t_1.

List of State Functions

The following are considered to be state functions in thermodynamics:

• Mass	• Pressure (P)
• Energy (E)	• Temperature (T)
○ Enthalpy (H)	• Volume (V)
○ Internal energy (U)	• Chemical composition
○ Gibbs free energy (G)	• Specific volume (v) or its reciprocal Density (ρ)
○ Helmholtz free energy (F)	
○ Exergy (B)	• Fugacity
• Entropy (S)	• Altitude
	• Particle number (n_i)

Atmospheric Thermodynamics

Atmospheric thermodynamics is the study of heat to work transformations (and the reverse) in the earth's atmospheric system *in relation to weather or climate*. Following the fundamental laws of classical thermodynamics, atmospheric thermodynamics studies such phenomena as properties of moist air, formation of clouds, atmospheric convection, boundary layer meteorology, and vertical stabilities in the atmosphere. Atmospheric thermodynamic diagrams are used as tools in the forecasting of storm development. Atmospheric thermodynamics forms a basis for cloud microphysics and convection parameterizations in numerical weather models, and is used in many climate considerations, including convective-equilibrium climate models.

Overview

The atmosphere is an example of a non-equilibrium system. Atmospheric thermodynamics focuses on water and its transformations. Areas of study include the law of energy conservation, the ideal gas law, specific heat capacities, adiabatic processes (in which entropy is conserved), and moist adiabatic processes. Most of tropospheric gases are treated as ideal gases and water vapor is considered as one of the most important trace components of air.

Advanced topics are phase transitions of water, homogeneous and inhomogeneous nucleation, effect of dissolved substances on cloud condensation, role of supersaturation on formation of ice crystals and cloud droplets. Considerations of moist air and cloud theories typically involve various temperatures, such as equivalent potential temperature, wet-bulb and virtual temperatures. Connected areas are energy, momentum, and mass transfer, turbulence interaction between air particles in clouds, convection, dynamics of tropical cyclones, and large scale dynamics of the atmosphere.

The major role of atmospheric thermodynamics is expressed in terms of adiabatic and diabatic forces acting on air parcels included in primitive equations of air motion either as grid resolved or subgrid parameterizations. These equations form a basis for the numerical weather and climate predictions.

History

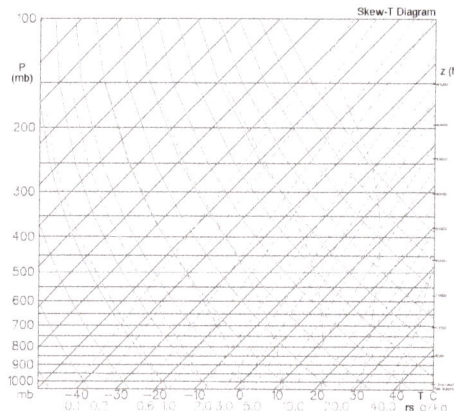

Thermodynamic diagram developed in the 19th century is still used to calculate quantities such as convective available potential energy or air stability.

In the early 19th century thermodynamicists such as Sadi Carnot, Rudolf Clausius, and Émile Clapeyron developed mathematical models on the dynamics of bodies fluids and vapors related to the combustion and pressure cycles of atmospheric steam engines; one example is the Clausius–Clapeyron equation. In 1873, thermodynamicist Willard Gibbs published "Graphical Methods in the Thermodynamics of Fluids."

These sorts of foundations naturally began to be applied towards the development of theoretical models of atmospheric thermodynamics which drew the attention of the best minds. Papers on atmospheric thermodynamics appeared in the 1860s that treated such topics as dry and moist adiabatic processes. In 1884 Heinrich Hertz devised first atmospheric thermodynamic diagram (emagram). Pseudo-adiabatic process was coined by von Bezold describing air as it is lifted, expands, cools, and eventually precipitates its water vapor; in 1888 he published voluminous work entitled "On the thermodynamics of the atmosphere".

In 1911 von Alfred Wegener published a book "Thermodynamik der Atmosphäre", Leipzig, J. A. Barth. From here the development of atmospheric thermodynamics as a branch of science began to take root. The term "atmospheric thermodynamics", itself, can be traced to Frank W. Verys 1919 publication: "The radiant properties of the earth from the standpoint of atmospheric thermodynamics" (Occasional scientific papers of the Westwood Astrophysical Observatory). By the late 1970s various textbooks on the subject began to appear. Today, atmospheric thermodynamics is an integral part of weather forecasting.

Chronology

- 1751 Charles Le Roy recognized dew point temperature as point of saturation of air

- 1782 Jacques Charles made hydrogen balloon flight measuring temperature and pressure in Paris

- 1784 Concept of variation of temperature with height was suggested

- 1801-1803 John Dalton developed his laws of pressures of vapours

- 1804 Joseph Louis Gay-Lussac made balloon ascent to study weather

- 1805 Pierre Simon Laplace developed his law of pressure variation with height

- 1841 James Pollard Espy publishes paper on convection theory of cyclone energy

- 1889 Hermann von Helmholtz and John William von Bezold used the concept of potential temperature, von Bezold used adiabatic lapse rate and pseudoadiabat

- 1893 Richard Asman constructs first aerological sonde (pressure-tempera-ture-humidity)

- 1894 John Wilhelm von Bezold used concept of equivalent temperature

- 1926 Sir Napier Shaw introduced tephigram

- 1933 Tor Bergeron published paper on "Physics of Clouds and Precipitation" describing precipitation from supercooled (due to condensational growth of ice crystals in presence of water drops)

- 1946 Vincent J. Schaeffer and Irving Langmuir performed the first cloud-seed-ing experiment

- 1986 K. Emanuel conceptualizes tropical cyclone as Carnot heat engine

Applications

Hadley Circulation

The Hadley Circulation can be considered as a heat engine. The Hadley circulation is identi-fied with rising of warm and moist air in the equatorial region with the descent of colder air in the subtropics corresponding to a thermally driven direct circulation, with consequent net production of kinetic energy. The thermodynamic efficiency of the Hadley system, con-sidered as a heat engine, has been relatively constant over the 1979~2010 period, averaging 2.6%. Over the same interval, the power generated by the Hadley regime has risen at an average rate of about 0.54 TW per yr; this reflects an increase in energy input to the system consistent with the observed trend in the tropical sea surface temperatures.

Tropical Cyclone Carnot Cycle

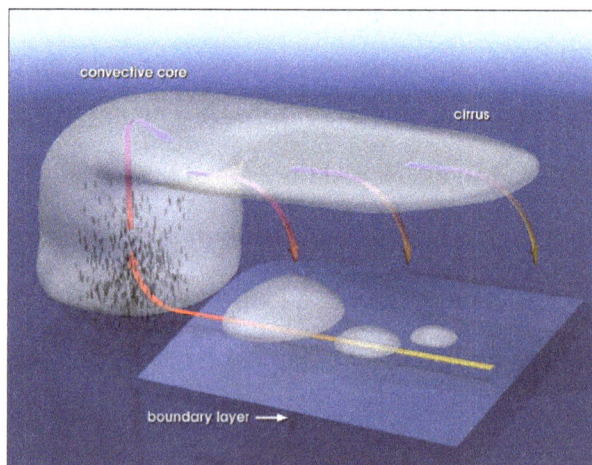

Air is being moistened as it travels toward convective system. Ascending motion in a deep convective core produces air expansion, cooling, and condensation. Upper level outflow visible as an anvil cloud is eventually descending conserving mass (rysunek - Robert Simmon).

The thermodynamic structure of the hurricane can be modelled as a heat engine running between sea temperature of about 300K and tropopause which has temperature of about 200K. Parcels of air traveling close to the surface take up moisture and warm, ascending air expands and cools releasing moisture (rain) during the condensation. The release of latent heat energy during the condensation provides mechanical energy for the hurricane. Both a decreasing temperature in the upper troposphere or an increasing temperature of the atmosphere close to the surface will increase the maximum winds observed in hurricanes. When applied to hurricane dynamics it defines a Carnot heat engine cycle and predicts maximum hurricane intensity.

Water Vapor and Global Climate Change

The Clausius–Clapeyron relation shows how the water-holding capacity of the atmosphere increases by about 8% per Celsius increase in temperature. (It does not directly depend on other parameters like the pressure or density.) This water-holding capacity, or "equilibrium vapor pressure", can be approximated using the August-Roche-Magnus formula

$$e_s(T) = 6.1094 \exp\left(\frac{17.625T}{T + 243.04}\right)$$

(where $e_s(T)$ is the equilibrium or saturation vapor pressure in hPa, and T is temperature in degrees Celsius). This shows that when atmospheric temperature increases (e.g., due to greenhouse gases) the absolute humidity should also increase exponentially (assuming a constant relative humidity). However, this purely thermodynamic argument is subject of considerable debate because convective processes might cause extensive drying due to increased areas of subsidence, efficiency of precipitation could be influenced by the intensity of convection, and because cloud formation is related to relative humidity.

Thermodynamic Diagrams

Thermodynamic diagrams are diagrams used to represent the thermodynamic states of a material (typically fluid) and the consequences of manipulating this material. For instance, a temperature-entropy diagram (T-S diagram) may be used to demonstrate the behavior of a fluid as it is changed by a compressor.

Overview

Especially in meteorology they are used to analyze the actual state of the atmosphere derived from the measurements of radiosondes, usually obtained with weather balloons. In such diagrams, temperature and humidity values (represented by the dew point) are displayed with respect to pressure. Thus the diagram gives at a first glance the actual atmospheric stratification and vertical water vapor distribution. Further analysis gives the actual base and top height of convective clouds or possible instabilities in the stratification.

By assuming the energy amount due to solar radiation it is possible to predict the 2 m (6.6 ft) temperature, humidity, and wind during the day, the development of the boundary layer of the atmosphere, the occurrence and development of clouds and the conditions for soaring flight during the day.

The main feature of thermodynamic diagrams is the equivalence between the area in the diagram and energy. When air changes pressure and temperature during a process and prescribes a closed curve within the diagram the area enclosed by this curve is proportional to the energy which has been gained or released by the air.

Types of Thermodynamic Diagrams

General purpose diagrams include:

- P-V diagram

- T-s diagram

- H-S (Mollier) diagram

- Psychrometric chart

- Cooling curve

- Indicator diagram

- Saturation vapor curve

- Thermodynamic Surface

Specific to weather services, there are mainly three different types of thermodynamic diagrams used:

- Skew-T log-P diagram

- Tephigram

- Emagram

These all are derived from the physical P-alpha-diagram which combines pressure and specific volume (alpha) as basic coordinates. The P-alpha-diagram shows a strong deformation of the grid for atmospheric conditions and is therefore not useful in atmospheric sciences. The three diagrams are constructed from the P-alpha-diagram by using appropriate coordinate transformations.

Not a thermodynamic diagram in a strict sense since it does not display the energy - area equivalence is the

- Stüve diagram

Due to its simpler construction it is preferred in education.

Characteristics

Thermodynamic diagrams usually show a net of five different lines:

- isobars = lines of constant pressure

- isotherms = lines of constant temperature

- dry adiabats = lines of constant potential temperature representing the temperature of a rising parcel of dry air

- saturated adiabats or pseudoadiabats = lines representing the temperature of a rising parcel saturated with water vapour

- mixing ratio = lines representing the dewpoint of a rising parcel

The lapse rate, dry adiabatic lapse rate (DALR) and moist adiabatic lapse rate (MALR), are obtained. With the help of these lines, parameters such as cloud condensation level, level of free convection, onset of cloud formation. etc. can be derived from the soundings.

Statistical Mechanics

Statistical mechanics is a branch of theoretical physics that studies, using probability theory, the average behavior of a mechanical system where the state of the system is uncertain.

A common use of statistical mechanics is in explaining the thermodynamic behavior of large systems. This branch of statistical mechanics which treats and extends classical thermodynamics is known as statistical thermodynamics or equilibrium statistical mechanics. Microscopic mechanical laws do not contain concepts such as temperature, heat, or entropy; however, statistical mechanics shows how these concepts arise from the natural uncertainty about the state of a system when that system is prepared in practice. The benefit of using statistical mechanics is that it provides exact methods to connect thermodynamic quantities (such as heat capacity) to microscopic behavior, whereas, in classical thermodynamics, the only available option would be to just measure and tabulate such quantities for various materials. Statistical mechanics also makes it possible to *extend* the laws of thermodynamics to cases which are not considered in classical thermodynamics, such as microscopic systems and other mechanical systems with few degrees of freedom.

Statistical mechanics also finds use outside equilibrium. An important subbranch known as non-equilibrium statistical mechanics deals with the issue of microscopically modelling the speed of irreversible processes that are driven by imbalances. Examples of such processes include chemical reactions or flows of particles and heat. Unlike with equilibrium, there is no exact formalism that applies to non-equilibrium statistical mechanics in general, and so this branch of statistical mechanics remains an active area of theoretical research.

Principles: Mechanics and Ensembles

In physics there are two types of mechanics usually examined: classical mechanics and quantum mechanics. For both types of mechanics, the standard mathematical approach is to consider two concepts:

1. The complete state of the mechanical system at a given time, mathematically encoded as a phase point (classical mechanics) or a pure quantum state vector (quantum mechanics).

2. An equation of motion which carries the state forward in time: Hamilton's equations (classical mechanics) or the time-dependent Schrödinger equation (quantum mechanics)

Using these two concepts, the state at any other time, past or future, can in principle be calculated. There is however a disconnection between these laws and everyday life experiences, as we do not find it necessary (nor even theoretically possible) to know exactly at a microscopic level the simultaneous positions and velocities of each molecule while carrying out processes at the human scale (for example, when performing a chemical reaction). Statistical mechanics fills this disconnection between the laws of mechanics and the practical experience of incomplete knowledge, by adding some uncertainty about which state the system is in.

Whereas ordinary mechanics only considers the behaviour of a single state, statistical mechanics introduces the statistical ensemble, which is a large collection of virtual, independent copies of the system in various states. The statistical ensemble is a probability distribution over all possible states of the system. In classical statistical mechanics, the ensemble is a probability distribution over phase points (as opposed to a single phase point in ordinary mechanics), usually represented as a distribution in a phase space with canonical coordinates. In quantum statistical mechanics, the ensemble is a probability distribution over pure states, and can be compactly summarized as a density matrix.

As is usual for probabilities, the ensemble can be interpreted in different ways:

*   an ensemble can be taken to represent the various possible states that a *single system* could be in (epistemic probability, a form of knowledge), or

- the members of the ensemble can be understood as the states of the systems in experiments repeated on independent systems which have been prepared in a similar but imperfectly controlled manner (empirical probability), in the limit of an infinite number of trials.

These two meanings are equivalent for many purposes, and will be used interchangeably in this article.

However the probability is interpreted, each state in the ensemble evolves over time according to the equation of motion. Thus, the ensemble itself (the probability distribution over states) also evolves, as the virtual systems in the ensemble continually leave one state and enter another. The ensemble evolution is given by the Liouville equation (classical mechanics) or the von Neumann equation (quantum mechanics). These equations are simply derived by the application of the mechanical equation of motion separately to each virtual system contained in the ensemble, with the probability of the virtual system being conserved over time as it evolves from state to state.

One special class of ensemble is those ensembles that do not evolve over time. These ensembles are known as *equilibrium ensembles* and their condition is known as *statistical equilibrium*. Statistical equilibrium occurs if, for each state in the ensemble, the ensemble also contains all of its future and past states with probabilities equal to the probability of being in that state. The study of equilibrium ensembles of isolated systems is the focus of statistical thermodynamics. Non-equilibrium statistical mechanics addresses the more general case of ensembles that change over time, and/or ensembles of non-isolated systems.

Statistical Thermodynamics

The primary goal of statistical thermodynamics (also known as equilibrium statistical mechanics) is to derive the classical thermodynamics of materials in terms of the properties of their constituent particles and the interactions between them. In other words, statistical thermodynamics provides a connection between the macroscopic properties of materials in thermodynamic equilibrium, and the microscopic behaviours and motions occurring inside the material.

Whereas statistical mechanics proper involves dynamics, here the attention is focussed on *statistical equilibrium* (steady state). Statistical equilibrium does not mean that the particles have stopped moving (mechanical equilibrium), rather, only that the ensemble is not evolving.

Fundamental Postulate

A sufficient (but not necessary) condition for statistical equilibrium with an isolated system is that the probability distribution is a function only of conserved properties (total energy, total particle numbers, etc.). There are many different equilibrium ensembles that can be considered, and only some of them correspond to thermodynamics. Additional postulates are necessary to motivate why the ensemble for a given system should have one form or another.

A common approach found in many textbooks is to take the *equal a priori probability postulate*. This postulate states that

> *For an isolated system with an exactly known energy and exactly known composition, the system can be found with* equal probability *in any microstate consistent with that knowledge.*

The equal a priori probability postulate therefore provides a motivation for the microcanonical ensemble described below. There are various arguments in favour of the equal a priori probability postulate:

- Ergodic hypothesis: An ergodic system is one that evolves over time to explore "all accessible" states: all those with the same energy and composition. In an ergodic system, the microcanonical ensemble is the only possible equilibrium ensemble with fixed energy. This approach has limited applicability, since most systems are not ergodic.

- Principle of indifference: In the absence of any further information, we can only assign equal probabilities to each compatible situation.

- Maximum information entropy: A more elaborate version of the principle of indifference states that the correct ensemble is the ensemble that is compatible with the known information and that has the largest Gibbs entropy (information entropy).

Other fundamental postulates for statistical mechanics have also been proposed.

Three Thermodynamic Ensembles

There are three equilibrium ensembles with a simple form that can be defined for any isolated system bounded inside a finite volume. These are the most often discussed ensembles in statistical thermodynamics. In the macroscopic limit (defined below) they all correspond to classical thermodynamics.

Microcanonical ensemble

> describes a system with a precisely given energy and fixed composition (precise number of particles). The microcanonical ensemble contains with equal probability each possible state that is consistent with that energy and composition.

Canonical ensemble

> describes a system of fixed composition that is in thermal equilibrium with a

heat bath of a precise temperature. The canonical ensemble contains states of varying energy but identical composition; the different states in the ensemble are accorded different probabilities depending on their total energy.

Grand canonical ensemble

describes a system with non-fixed composition (uncertain particle numbers) that is in thermal and chemical equilibrium with a thermodynamic reservoir. The reservoir has a precise temperature, and precise chemical potentials for various types of particle. The grand canonical ensemble contains states of varying energy and varying numbers of particles; the different states in the ensemble are accorded different probabilities depending on their total energy and total particle numbers.

For systems containing many particles (the thermodynamic limit), all three of the ensembles listed above tend to give identical behaviour. It is then simply a matter of mathematical convenience which ensemble is used.

Important cases where the thermodynamic ensembles *do not* give identical results include:

- Microscopic systems.

- Large systems at a phase transition.

- Large systems with long-range interactions.

In these cases the correct thermodynamic ensemble must be chosen as there are observable differences between these ensembles not just in the size of fluctuations, but also in average quantities such as the distribution of particles. The correct ensemble is that which corresponds to the way the system has been prepared and characterized—in other words, the ensemble that reflects the knowledge about that system.

Calculation Methods

Once the characteristic state function for an ensemble has been calculated for a given system, that system is 'solved' (macroscopic observables can be extracted from the characteristic state function). Calculating the characteristic state function of a thermodynamic ensemble is not necessarily a simple task, however, since it involves considering every possible state of the system. While some hypothetical systems have been exactly solved, the most general (and realistic) case is too complex for exact solution. Various approaches exist to approximate the true ensemble and allow calculation of average quantities.

Exact

There are some cases which allow exact solutions.

- For very small microscopic systems, the ensembles can be directly computed by simply enumerating over all possible states of the system (using exact diagonalization in quantum mechanics, or integral over all phase space in classical mechanics).

- Some large systems consist of many separable microscopic systems, and each of the subsystems can be analysed independently. Notably, idealized gases of non-interacting particles have this property, allowing exact derivations of Maxwell–Boltzmann statistics, Fermi–Dirac statistics, and Bose–Einstein statistics.

- A few large systems with interaction have been solved. By the use of subtle mathematical techniques, exact solutions have been found for a few toy models. Some examples include the Bethe ansatz, square-lattice Ising model in zero field, hard hexagon model.

Monte Carlo

One approximate approach that is particularly well suited to computers is the Monte Carlo method, which examines just a few of the possible states of the system, with the states chosen randomly (with a fair weight). As long as these states form a representative sample of the whole set of states of the system, the approximate characteristic function is obtained. As more and more random samples are included, the errors are reduced to an arbitrarily low level.

- The Metropolis–Hastings algorithm is a classic Monte Carlo method which was initially used to sample the canonical ensemble.

- Path integral Monte Carlo, also used to sample the canonical ensemble.

Other

- For rarefied non-ideal gases, approaches such as the cluster expansion use perturbation theory to include the effect of weak interactions, leading to a virial expansion.

- For dense fluids, another approximate approach is based on reduced distribution functions, in particular the radial distribution function.

- Molecular dynamics computer simulations can be used to calculate microcanonical ensemble averages, in ergodic systems. With the inclusion of a connection to a stochastic heat bath, they can also model canonical and grand canonical conditions.

- Mixed methods involving non-equilibrium statistical mechanical results may be useful.

Non-equilibrium Statistical Mechanics

There are many physical phenomena of interest that involve quasi-thermodynamic processes out of equilibrium, for example:

- heat transport by the internal motions in a material, driven by a temperature imbalance,

- electric currents carried by the motion of charges in a conductor, driven by a voltage imbalance,

- spontaneous chemical reactions driven by a decrease in free energy,

- friction, dissipation, quantum decoherence,

- systems being pumped by external forces (optical pumping, etc.),

- and irreversible processes in general.

All of these processes occur over time with characteristic rates, and these rates are of importance for engineering. The field of non-equilibrium statistical mechanics is concerned with understanding these non-equilibrium processes at the microscopic level. (Statistical thermodynamics can only be used to calculate the final result, after the external imbalances have been removed and the ensemble has settled back down to equilibrium.)

In principle, non-equilibrium statistical mechanics could be mathematically exact: ensembles for an isolated system evolve over time according to deterministic equations such as Liouville's equation or its quantum equivalent, the von Neumann equation. These equations are the result of applying the mechanical equations of motion independently to each state in the ensemble. Unfortunately, these ensemble evolution equations inherit much of the complexity of the underlying mechanical motion, and so exact solutions are very difficult to obtain. Moreover, the ensemble evolution equations are fully reversible and do not destroy information (the ensemble's Gibbs entropy is preserved). In order to make headway in modelling irreversible processes, it is necessary to consider additional factors besides probability and reversible mechanics.

Non-equilibrium mechanics is therefore an active area of theoretical research as the range of validity of these additional assumptions continues to be explored. A few approaches are described in the following subsections.

Stochastic Methods

One approach to non-equilibrium statistical mechanics is to incorporate stochastic (random) behaviour into the system. Stochastic behaviour destroys information contained in the ensemble. While this is technically inaccurate (aside from hypothetical situations involving black holes, a system cannot in itself cause loss of information), the randomness is added to reflect that information of interest becomes converted over time into subtle correlations within the system, or to correlations between the system and environment. These correlations appear as chaotic or pseudorandom influences on the variables of interest. By replacing these correlations with randomness proper, the calculations can be made much easier.

- *Boltzmann transport equation*: An early form of stochastic mechanics appeared even before the term "statistical mechanics" had been coined, in studies of kinetic theory. James Clerk Maxwell had demonstrated that molecular collisions would lead to apparently chaotic motion inside a gas. Ludwig Boltzmann subsequently showed that, by taking this molecular chaos for granted as a complete randomization, the motions of particles in a gas would follow a simple Boltzmann transport equation that would rapidly restore a gas to an equilibrium state.

  The Boltzmann transport equation and related approaches are important tools in non-equilibrium statistical mechanics due to their extreme simplicity. These approximations work well in systems where the "interesting" information is immediately (after just one collision) scrambled up into subtle correlations, which essentially restricts them to rarefied gases. The Boltzmann transport equation has been found to be very useful in simulations of electron transport in lightly doped semiconductors (in transistors), where the electrons are indeed analogous to a rarefied gas.

  A quantum technique related in theme is the random phase approximation.

- *BBGKY hierarchy*: In liquids and dense gases, it is not valid to immediately discard the correlations between particles after one collision. The BBGKY hierarchy (Bogoliubov–Born–Green–Kirkwood–Yvon hierarchy) gives a method for deriving Boltzmann-type equations but also extending them beyond the dilute gas case, to include correlations after a few collisions.

- *Keldysh formalism* (a.k.a. NEGF—non-equilibrium Green functions): A quantum approach to including stochastic dynamics is found in the Keldysh formalism. This approach often used in electronic quantum transport calculations.

Near-equilibrium Methods

Another important class of non-equilibrium statistical mechanical models deals with systems that are only very slightly perturbed from equilibrium. With very small perturbations, the response can be analysed in linear response theory. A remarkable result, as formalized by the fluctuation-dissipation theorem, is that the response of a system when near equilibrium is precisely related to the fluctuations that occur when the system is in total equilibrium. Essentially, a system that is slightly away from equilibrium—whether put there by external forces or by fluctuations—relaxes towards equilibrium in the same way, since the system cannot tell the difference or "know" how it came to be away from equilibrium.

This provides an indirect avenue for obtaining numbers such as ohmic conductivity and thermal conductivity by extracting results from equilibrium statistical mechanics.

Since equilibrium statistical mechanics is mathematically well defined and (in some cases) more amenable for calculations, the fluctuation-dissipation connection can be a convenient shortcut for calculations in near-equilibrium statistical mechanics.

A few of the theoretical tools used to make this connection include:

- Fluctuation–dissipation theorem
- Onsager reciprocal relations
- Green–Kubo relations
- Landauer–Büttiker formalism
- Mori–Zwanzig formalism

Hybrid Methods

An advanced approach uses a combination of stochastic methods and linear response theory. As an example, one approach to compute quantum coherence effects (weak localization, conductance fluctuations) in the conductance of an electronic system is the use of the Green-Kubo relations, with the inclusion of stochastic dephasing by interactions between various electrons by use of the Keldysh method.

Applications Outside Thermodynamics

The ensemble formalism also can be used to analyze general mechanical systems with uncertainty in knowledge about the state of a system. Ensembles are also used in:

- propagation of uncertainty over time,
- regression analysis of gravitational orbits,
- ensemble forecasting of weather,
- dynamics of neural networks,
- bounded-rational potential games in game theory and economics.

History

In 1738, Swiss physicist and mathematician Daniel Bernoulli published *Hydrodynamica* which laid the basis for the kinetic theory of gases. In this work, Bernoulli posited the argument, still used to this day, that gases consist of great numbers of molecules moving in all directions, that their impact on a surface causes the gas pressure that we feel, and that what we experience as heat is simply the kinetic energy of their motion.

In 1859, after reading a paper on the diffusion of molecules by Rudolf Clausius, Scottish physicist James Clerk Maxwell formulated the Maxwell distribution of molecular velocities, which gave the proportion of molecules having a certain velocity in a specific range. This was the first-ever statistical law in physics. Five years later, in 1864, Ludwig Boltzmann, a young student in Vienna, came across Maxwell's paper and spent much of his life developing the subject further.

Statistical mechanics proper was initiated in the 1870s with the work of Boltzmann, much of which was collectively published in his 1896 *Lectures on Gas Theory*. Boltzmann's original papers on the statistical interpretation of thermodynamics, the H-theorem, transport theory, thermal equilibrium, the equation of state of gases, and similar subjects, occupy about 2,000 pages in the proceedings of the Vienna Academy and other societies. Boltzmann introduced the concept of an equilibrium statistical ensemble and also investigated for the first time non-equilibrium statistical mechanics, with his *H*-theorem.

The term "statistical mechanics" was coined by the American mathematical physicist J. Willard Gibbs in 1884. "Probabilistic mechanics" might today seem a more appropriate term, but "statistical mechanics" is firmly entrenched. Shortly before his death, Gibbs published in 1902 *Elementary Principles in Statistical Mechanics*, a book which formalized statistical mechanics as a fully general approach to address all mechanical systems—macroscopic or microscopic, gaseous or non-gaseous. Gibbs' methods were initially derived in the framework classical mechanics, however they were of such generality that they were found to adapt easily to the later quantum mechanics, and still form the foundation of statistical mechanics to this day.

References

- van Gool, W.; Bruggink, J.J.C. (Eds) (1985). Energy and time in the economic and physical sciences. North-Holland. pp. 41–56. ISBN 0444877487.

- Adkins, C.J. (1968/1975). Equilibrium Thermodynamics, second edition, McGraw-Hill, London, ISBN 0-07-084057-1.

- Bailyn, M. (1994). A Survey of Thermodynamics, American Institute of Physics Press, New York, ISBN 0-88318-797-3.

- Callen, H.B. (1960/1985), Thermodynamics and an Introduction to Thermostatistics, (first edition 1960), second edition 1985, John Wiley & Sons, New York, ISBN 0-471-86256-8.

- Münster, A. (1970), Classical Thermodynamics, translated by E.S. Halberstadt, Wiley–Interscience, London, ISBN 0-471-62430-6.

- Tschoegl, N.W. (2000). Fundamentals of Equilibrium and Steady-State Thermodynamics, Elsevier, Amsterdam, ISBN 0-444-50426-5.

- Clausius, Rudolf (1850). On the Motive Power of Heat, and on the Laws which can be deduced from it for the Theory of Heat. Poggendorff's Annalen der Physik, LXXIX (Dover Reprint). ISBN 0-486-59065-8.

- Clark, John, O.E. (2004). The Essential Dictionary of Science. Barnes & Noble Books. ISBN 0-7607-4616-8. OCLC 58732844.

- Van Ness, H.C. (1983) [1969]. Understanding Thermodynamics. Dover Publications, Inc. ISBN 9780486632773. OCLC 8846081.

- Smith, J.M.; Van Ness, H.C.; Abbott, M.M. (2005). Introduction to Chemical Engineering Thermodynamics. McGraw Hill. ISBN 0-07-310445-0. OCLC 56491111.

- Cengel, Yunus A.; Boles, Michael A. (2005). Thermodynamics - an Engineering Approach. McGraw-Hill. ISBN 0-07-310768-9.

- Gibbs, Willard (1993). The Scientific Papers of J. Willard Gibbs, Volume One: Thermodynamics. Ox Bow Press. ISBN 0-918024-77-3. OCLC 27974820.

- Guggenheim, E.A. (1985). Thermodynamics. An Advanced Treatment for Chemists and Physicists, seventh edition, North Holland, Amsterdam, ISBN 0-444-86951-4.

- Kittel, C. Kroemer, H. (1980). Thermal Physics, second edition, W.H. Freeman, San Francisco, ISBN 0-7167-1088-9.

- Lebon, G., Jou, D., Casas-Vázquez, J. (2008). Understanding Non-equilibrium Thermodynamics. Foundations, Applications, Frontiers, Springer, Berlin, ISBN 978-3-540-74252-4.

- Chris Vuille; Serway, Raymond A.; Faughn, Jerry S. (2009). College physics. Belmont, CA: Brooks/Cole, Cengage Learning. p. 355. ISBN 0-495-38693-6.

- Glansdorff, P., Prigogine, I. (1971). Thermodynamic Theory of Structure, Stability and Fluctuations, Wiley-Interscience, London, ISBN 0-471-30280-5.

- Serrin, J. (1986). Chapter 1, 'An Outline of Thermodynamical Structure', pages 3-32, in New Perspectives in Thermodynamics, edited by J. Serrin, Springer, Berlin, ISBN 3-540-15931-2.

- Adkins, C.J. (1968/1983). Equilibrium Thermodynamics, (first edition 1968), third edition 1983, Cambridge University Press, ISBN 0-521-25445-0, pp. 18–20.

- Bailyn, M. (1994). A Survey of Thermodynamics, American Institute of Physics Press, New York, ISBN 0-88318-797-3, p. 26.

- Kittel, C. Kroemer, H. (1980). Thermal Physics, (first edition by Kittel alone 1969), second edition, W. H. Freeman, San Francisco, ISBN 0-7167-1088-9, pp. 49, 227.

- Tro, N. J. (2008). Chemistry. A Molecular Approach, Pearson/Prentice Hall, Upper Saddle River NJ, ISBN 0-13-100065-9, p. 246.

5

Laws of Thermodynamics

Thermodynamics is basically based on four sets of laws. The most prominent law in thermodynamics is the zeroth law of thermodynamics. The additional laws are the first law of thermodynamics, second law of thermodynamics and third law of thermodynamics. This chapter discusses the thermodynamics in a critical manner providing key analysis to the subject matter.

Laws of Thermodynamics

The four laws of thermodynamics define fundamental physical quantities (temperature, energy, and entropy) that characterize thermodynamic systems at thermal equilibrium. The laws describe how these quantities behave under various circumstances, and forbid certain phenomena (such as perpetual motion).

The four laws of thermodynamics are:

- Zeroth law of thermodynamics: If two systems are in thermal equilibrium with a third system, they are in thermal equilibrium with each other. This law helps define the notion of temperature.

- First law of thermodynamics: When energy passes, as work, as heat, or with matter, into or out from a system, the system's internal energy changes in accord with the law of conservation of energy. Equivalently, perpetual motion machines of the first kind are impossible.

- Second law of thermodynamics: In a natural thermodynamic process, the sum of the entropies of the interacting thermodynamic systems increases. Equivalently, perpetual motion machines of the second kind are impossible.

- Third law of thermodynamics: The entropy of a system approaches a constant value as the temperature approaches absolute zero. With the exception of non-crystalline solids (glasses) the entropy of a system at absolute zero is typically close to zero, and is equal to the logarithm of the product of the quantum ground states.

There have been suggestions of additional laws, but none of them achieves the generality of the four accepted laws, and they are not mentioned in standard textbooks.

The laws of thermodynamics are important fundamental laws in physics and they are applicable in other natural sciences.

Zeroth Law

The zeroth law of thermodynamics may be stated in the following form:

If two systems are both in thermal equilibrium with a third system then they are in thermal equilibrium with each other.

The law is intended to allow the existence of an empirical parameter, the temperature, as a property of a system such that systems in thermal equilibrium with each other have the same temperature. The law as stated here is compatible with the use of a particular physical body, for example a mass of gas, to match temperatures of other bodies, but does not justify regarding temperature as a quantity that can be measured on a scale of real numbers.

Though this version of the law is one of the more commonly stated, it is only one of a diversity of statements that are labeled as "the zeroth law" by competent writers. Some statements go further so as to supply the important physical fact that temperature is one-dimensional, that one can conceptually arrange bodies in real number sequence from colder to hotter. Perhaps there exists no unique "best possible statement" of the "zeroth law", because there is in the literature a range of formulations of the principles of thermodynamics, each of which call for their respectively appropriate versions of the law.

Although these concepts of temperature and of thermal equilibrium are fundamental to thermodynamics and were clearly stated in the nineteenth century, the desire to ex-plicitly number the above law was not widely felt until Fowler and Guggenheim did so in the 1930s, long after the first, second, and third law were already widely understood and recognized. Hence it was numbered the zeroth law. The importance of the law as a foundation to the earlier laws is that it allows the definition of temperature in a non-cir-cular way without reference to entropy, its conjugate variable. Such a temperature defi-nition is said to be 'empirical'.

First Law

The first law of thermodynamics may be stated in several ways :

> The increase in internal energy of a closed system is equal to total of the energy added to the system. In particular, if the energy entering the system is supplied as heat and energy leaves the system as work, the heat is accounted as positive and the work is accounted as negative.
>
> $$\Delta U_{system} = Q - W$$
>
> In the case of a thermodynamic cycle of a closed system, which returns to its original state, the heat Q_{in} supplied to the system in one stage of the cycle, minus the heat Q_{out} removed from it in another stage of the cycle, plus the work added to the system W_{in} equals the work that leaves the system W_{out}.

$$\Delta U_{system(full\,cycle)} = 0$$

hence, for a full cycle,

$$Q = Q_{in} - Q_{out} + W_{in} - W_{out} = W_{net}$$

For the particular case of a thermally isolated system (adiabatically isolated), the change of the internal energy of an adiabatically isolated system can only be the result of the work added to the system, because the adiabatic assumption is: $Q = 0$.

$$\Delta U_{system} = U_{final} - U_{initial} = W_{in} - W_{out}$$

More specifically, the First Law encompasses several principles:

- The law of conservation of energy.

  This states that energy can be neither created nor destroyed. However, energy can change forms, and energy can flow from one place to another. A particular consequence of the law of conservation of energy is that the total energy of an isolated system does not change.

- The concept of internal energy and its relationship to temperature.

  If a system has a definite temperature, then its total energy has three distinguishable components. If the system is in motion as a whole, it has kinetic energy. If the system as a whole is in an externally imposed force field (e.g. gravity), it has potential energy relative to some reference point in space. Finally, it has internal energy, which is a fundamental quantity of thermodynamics. The establishment of the concept of internal energy distinguishes the first law of thermodynamics from the more general law of conservation of energy.

$$E_{total} = KE_{system} + PE_{system} + U_{system}$$

  The internal energy of a substance can be explained as the sum of the diverse kinetic energies of the erratic microscopic motions of its constituent atoms, and of the potential energy of interactions between them. Those microscopic energy terms are collectively called the substance's internal energy (U), and are accounted for by macroscopic thermodynamic property. The total of the kinetic energies of microscopic motions of the constituent atoms increases as the system's temperature increases; this assumes no other interactions at the microscopic level of the system such as chemical reactions, potential energy of constituent atoms with respect to each other.

- Work is a process of transferring energy to or from a system in ways that can be described by macroscopic mechanical forces exerted by factors in the surroundings, outside the system. Examples are an externally driven shaft agitating a stirrer within the system, or an externally imposed electric field that polarizes the material of the system, or a piston that compresses the system. Unless otherwise stated, it is customary to treat work as occurring without its dissipation to the surroundings. Practically speaking, in all natural process, some of the work is dissipated by internal friction or viscosity. The work done by the system can come from its overall kinetic energy, from its overall potential energy, or from its internal energy.

For example, when a machine (not a part of the system) lifts a system upwards, some energy is transferred from the machine to the system. The system's energy increases as work is done on the system and in this particular case, the energy increase of the system is manifested as an increase in the system's gravitational potential energy. Work added to the system increases the Potential Energy of the system:

$$W = \Delta PE_{system}$$

Or in general, the energy added to the system in the form of work can be partitioned to kinetic, potential or internal energy forms:

$$W = \Delta KE_{system} + \Delta PE_{system} + \Delta U_{system}$$

- When matter is transferred into a system, that masses' associated internal energy and potential energy are transferred with it.

$$\left(u \Delta M \right)_{in} = \Delta U_{system}$$

where u denotes the internal energy per unit mass of the transferred matter, as measured while in the surroundings; and ΔM denotes the amount of transferred mass.

- The flow of heat is a form of energy transfer.

Heating is a natural process of moving energy to or from a system other than by work or the transfer of matter. Direct passage of heat is only from a hotter to a colder system.

If the system has rigid walls that are impermeable to matter, and consequently energy cannot be transferred as work into or out from the system, and no external long-range force field affects it that could change its internal energy, then the internal energy can only be changed by the transfer of energy as heat:

$$\Delta U_{system} = Q$$

where Q denotes the amount of energy transferred into the system as heat.

Combining these principles leads to one traditional statement of the first law of thermodynamics: it is not possible to construct a machine which will perpetually output work without an equal amount of energy input to that machine. Or more briefly, a perpetual motion machine of the first kind is impossible.

Second Law

The second law of thermodynamics indicates the irreversibility of natural processes, and, in many cases, the tendency of natural processes to lead towards spatial homogeneity of matter and energy, and especially of temperature. It can be formulated in a variety of interesting and important ways.

It implies the existence of a quantity called the entropy of a thermodynamic system. In terms of this quantity it implies that

When two initially isolated systems in separate but nearby regions of space, each in thermodynamic equilibrium with itself but not necessarily with each other, are then allowed to interact, they will eventually reach a mutual thermodynamic equilibrium. The sum of the entropies of the initially isolated systems is less than or equal to the total entropy of the final combination. Equality occurs just when the two original systems have all their respective intensive variables (temperature, pressure) equal; then the final system also has the same values.

This statement of the second law is founded on the assumption, that in classical thermodynamics, the entropy of a system is defined only when it has reached internal thermodynamic equilibrium (thermodynamic equilibrium with itself).

The second law is applicable to a wide variety of processes, reversible and irreversible. All natural processes are irreversible. Reversible processes are a useful and convenient theoretical fiction, but do not occur in nature.

A prime example of irreversibility is in the transfer of heat by conduction or radiation. It was known long before the discovery of the notion of entropy that when two bodies initially of different temperatures come into thermal connection, then heat always flows from the hotter body to the colder one.

The second law tells also about kinds of irreversibility other than heat transfer, for example those of friction and viscosity, and those of chemical reactions. The notion of entropy is needed to provide that wider scope of the law.

According to the second law of thermodynamics, in a theoretical and fictive reversible heat transfer, an element of heat transferred, δQ, is the product of the temperature (T), both of the system and of the sources or destination of the heat, with the increment (dS) of the system's conjugate variable, its entropy (S)

$$\delta Q = T\,dS.$$

Entropy may also be viewed as a physical measure of the lack of physical information about the microscopic details of the motion and configuration of a system, when only the macroscopic states are known. The law asserts that for two given macroscopically specified states of a system, there is a quantity called the difference of information entropy between them. This information entropy difference defines how much additional microscopic physical information is needed to specify one of the macroscopically specified states, given the macroscopic specification of the other - often a conveniently chosen reference state which may be presupposed to exist rather than explicitly stated. A final condition of a natural process always contains microscopically specifiable effects which are not fully and exactly predictable from the macroscopic specification of the initial condition of the process. This is why entropy increases in natural processes - the increase tells how much extra microscopic information is needed to distinguish the final macroscopically specified state from the initial macroscopically specified state.

Third Law

The third law of thermodynamics is sometimes stated as follows:

> *The entropy of a perfect crystal of any pure substance approaches zero as the temperature approaches absolute zero.*

At zero temperature the system must be in a state with the minimum thermal energy. This statement holds true if the perfect crystal has only one state with minimum energy. Entropy is related to the number of possible microstates according to:

$$S = k_B \ln \Omega$$

Where S is the entropy of the system, k_B Boltzmann's constant, and Ω the number of microstates (e.g. possible configurations of atoms). At absolute zero there is only 1 microstate possible ($\Omega=1$ as all the atoms are identical for a pure substance and as a result all orders are identical as there is only one combination) and $\ln(1) = 0$.

A more general form of the third law that applies to a system such as a glass that may have more than one minimum microscopically distinct energy state, or may have a microscopically distinct state that is "frozen in" though not a strictly minimum energy state and not strictly speaking a state of thermodynamic equilibrium, at absolute zero temperature:

> *The entropy of a system approaches a constant value as the temperature approaches zero.*

The constant value (not necessarily zero) is called the residual entropy of the system.

History

Circa 1797, Count Rumford (born Benjamin Thompson) showed that endless mechanical action can generate indefinitely large amounts of heat from a fixed amount of working substance thus challenging the caloric theory of heat, which held that there would be a finite amount of caloric heat/energy in a fixed amount of working substance. The first established thermodynamic principle, which eventually became the second law of thermodynamics, was formulated by Sadi Carnot in 1824. By 1860, as formalized in the works of those such as Rudolf Clausius and William Thomson, two established principles of thermodynamics had evolved, the first principle and the second principle, later restated as thermodynamic laws. By 1873, for example, thermodynamicist Josiah Willard Gibbs, in his memoir *Graphical Methods in the Thermodynamics of Fluids*, clearly stated the first two absolute laws of thermodynamics. Some textbooks throughout the 20th century have numbered the laws differently. In some fields removed from chemistry, the second law was considered to deal with the efficiency of heat engines only, whereas what was called the third law dealt with entropy increases. Directly defining zero points for entropy calculations was not considered to be a law. Gradually, this separation was combined into the second law and the modern third law was widely adopted.

Four Laws of Thermodynamics

Zeroth Law of Thermodynamics

The zeroth law of thermodynamics states that if two thermodynamic systems are each in thermal equilibrium with a third, then they are in thermal equilibrium with each other.

Two systems are said to be in the relation of thermal equilibrium if they are linked by a wall permeable only to heat and they do not change over time. As a convenience of language, systems are sometimes also said to be in a relation of thermal equilibrium if they are not linked so as to be able to transfer heat to each other, but would not do so if they were connected by a wall permeable only to heat. Thermal equilibrium between two systems is a transitive relation.

The physical meaning of the law was expressed by Maxwell in the words: "All heat is of the same kind". For this reason, another statement of the law is "All diathermal walls are equivalent".

The law is important for the mathematical formulation of thermodynamics, which needs the assertion that the relation of thermal equilibrium is an equivalence relation. This information is needed for a mathematical definition of temperature that will agree with the physical existence of valid thermometers.

Zeroth Law as Equivalence Relation

A thermodynamic system is by definition in its own state of internal thermodynamic equilibrium, that is to say, there is no change in its observable state (i.e. macrostate) over time and no flows occur in it. One precise statement of the zeroth law is that the relation of thermal equilibrium is an equivalence relation on pairs of thermodynamic systems. In other words, the set of all systems each in its own state of internal thermodynamic equilibrium may be divided into subsets in which every system belongs to one and only one subset, and is in thermal equilibrium with every other member of that subset, and is not in thermal equilibrium with a member of any other subset. This means that a unique "tag" can be assigned to every system, and if the "tags" of two systems are the same, they are in thermal equilibrium with each other, and if different, they are not. This property is used to justify the use of empirical temperature as a tagging system. Empirical temperature provides further relations of thermally equilibrated systems, such as order and continuity with regard to "hotness" or "coldness", but these are not implied by the standard statement of the zeroth law.

If it is defined that a thermodynamic system is in thermal equilibrium with itself (i.e., thermal equilibrium is reflexive), then the zeroth law may be stated as follows:

If a body A, *be in thermal equilibrium with two other bodies,* B *and* C, *then* B *and* C *are in thermal equilibrium with one another.*

This statement asserts that thermal equilibrium is a left-Euclidean relation between thermodynamic systems. If we also define that every thermodynamic system is in thermal equilibrium with itself, then thermal equilibrium is also a reflexive relation. Binary relations that are both reflexive and Euclidean are equivalence relations. Thus, again implicitly assuming reflexivity, the zeroth law is therefore often expressed as a right-Euclidean statement:

If two systems are in thermal equilibrium with a third system, then they are in thermal equilibrium with each other.

One consequence of an equivalence relationship is that the equilibrium relationship is symmetric: If *A* is in thermal equilibrium with *B*, then *B* is in thermal equilibrium with *A*. Thus we may say that two systems are in thermal equilibrium with each other, or that they are in mutual equilibrium. Another consequence of equivalence is that thermal equilibrium is a transitive relationship and is occasionally expressed as such:

If A *is in thermal equilibrium with* B *and if* B *is in thermal equilibrium with* C, *then* A *is in thermal equilibrium with* C .

A reflexive, transitive relationship does not guarantee an equivalence relationship. In order for the above statement to be true, *both* reflexivity *and* symmetry must be implicitly assumed.

It is the Euclidean relationships which apply directly to thermometry. An ideal thermometer is a thermometer which does not measurably change the state of the system it is measuring. Assuming that the unchanging reading of an ideal thermometer is a valid "tagging" system for the equivalence classes of a set of equilibrated thermodynamic systems, then if a thermometer gives the same reading for two systems, those two systems are in thermal equilibrium, and if we thermally connect the two systems, there will be no subsequent change in the state of either one. If the readings are different, then thermally connecting the two systems will cause a change in the states of both systems and when the change is complete, they will both yield the same thermometer reading. The zeroth law provides no information regarding this final reading.

Foundation of Temperature

The zeroth law establishes thermal equilibrium as an equivalence relationship. An equivalence relationship on a set (such as the set of all systems each in its own state of internal thermodynamic equilibrium) divides that set into a collection of distinct subsets ("disjoint subsets") where any member of the set is a member of one and only one such subset. In the case of the zeroth law, these subsets consist of systems which are in mutual equilibrium. This partitioning allows any member of the subset to be uniquely "tagged" with a label identifying the subset to which it belongs. Although the labeling may be quite arbitrary, temperature is just such a labeling process which uses the real number system for tagging. The zeroth law justifies the use of suitable thermodynamic systems as thermometers to provide such a labeling, which yield any number of possible empirical temperature scales, and justifies the use of the second law of thermodynamics to provide an absolute, or thermodynamic temperature scale. Such temperature scales bring additional continuity and ordering (i.e., "hot" and "cold") properties to the concept of temperature.

In the space of thermodynamic parameters, zones of constant temperature form a surface, that provides a natural order of nearby surfaces. One may therefore construct a global temperature function that provides a continuous ordering of states. The dimensionality of a surface of constant temperature is one less than the number of thermodynamic parameters, thus, for an ideal gas described with three thermodynamic parameters P, V and N, it is a two-dimensional surface.

For example, if two systems of ideal gases are in equilibrium, then $\frac{P_1 V_1}{N_1} = \frac{P_2 V_2}{N_2}$ where P_i is the pressure in the ith system, V_i is the volume, and N_i is the amount (in moles, or simply the number of atoms) of gas.

The surface $\frac{PV}{N}$ = constant defines surfaces of equal thermodynamic temperature, and one may label defining T so that $\frac{PV}{N} = RT$, where R is some constant. These systems can now be used as a thermometer to calibrate other systems. Such systems are known as "ideal gas thermometers".

In a sense, focused on in the zeroth law, there is only one kind of diathermal wall or one kind of heat, as expressed by Maxwell's dictum that "All heat is of the same kind". But in another sense, heat is transferred in different ranks, as expressed by Sommerfeld's dictum "Thermodynamics investigates the conditions that govern the transformation of heat into work. It teaches us to recognize temperature as the measure of the work-value of heat. Heat of higher temperature is richer, is capable of doing more work. Work may be regarded as heat of an infinitely high temperature, as unconditionally available heat." This is why temperature is the particular variable indicated by the zeroth law's statement of equivalence.

Physical Meaning of the Usual Statement of the Zeroth Law

The present article states the zeroth law as it is often summarized in textbooks. Nevertheless, this usual statement perhaps does not explicitly convey the full physical meaning that underlies it. The underlying physical meaning was perhaps first clarified by Maxwell in his 1871 textbook.

In Carathéodory's (1909) theory, it is postulated that there exist walls "permeable only to heat", though heat is not explicitly defined in that paper. This postulate is a physical postulate of existence. It does not, however, as worded just previously, say that there is only one kind of heat. This paper of Carathéodory states as proviso 4 of its account of such walls: "Whenever each of the systems S_1 and S_2 is made to reach equilibrium with a third system S_3 under identical conditions, systems S_1 and S_2 are in mutual equilibrium". It is the function of this statement in the paper, not there labeled as the zeroth law, to provide not only for the existence of transfer of energy other than by work or transfer of matter, but further to provide that such transfer is unique in the sense that there is only one kind of such wall, and one kind of such transfer. This is signaled in the postulate of this paper of Carathéodory that precisely one non-deformation variable is needed to complete the specification of a thermodynamic state, beyond the necessary deformation variables, which are not restricted in number. It is therefore not exactly clear what Carathéodory means when in the introduction of this paper he writes "*It is possible to develop the whole theory without assuming the existence of heat, that is of a quantity that is of a different nature from the normal mechanical quantities.*"

Maxwell (1871) discusses at some length ideas which he summarizes by the words "All heat is of the same kind". Modern theorists sometimes express this idea by postulating the existence of a unique one-dimensional *hotness manifold*, into which every proper temperature scale has a monotonic mapping. This may be expressed by the statement that there is only one kind of temperature, regardless of the variety of scales in which it is expressed. Another modern expression of this idea is that "All diathermal walls are equivalent". This might also be expressed by saying that there is precisely one kind of non-mechanical, non-matter-transferring contact equilibrium between thermodynamic systems.

These ideas may be regarded as helping to clarify the physical meaning of the usual statement of the zeroth law of thermodynamics. It is the opinion of Lieb and Yngvason (1999) that the derivation from statistical mechanics of the law of entropy increase is a goal that has so far eluded the deepest thinkers. Thus the idea remains open to consideration that the existence of heat and temperature are needed as coherent primitive concepts for thermodynamics, as expressed, for example, by Maxwell and Planck. On the other hand, Planck in 1926 clarified how the second law can be stated without reference to heat or temperature, by referring to the irreversible and universal nature of friction in natural thermodynamic processes.

History

According to Arnold Sommerfeld, Ralph H. Fowler invented the title 'the zeroth law of thermodynamics' when he was discussing the 1935 text of Saha and Srivastava. They write on page 1 that "every physical quantity must be measurable in numerical terms". They presume that temperature is a physical quantity and then deduce the statement "If a body A is in temperature equilibrium with two bodies B and C, then B and C themselves will be in temperature equilibrium with each other". They then in a self-standing paragraph italicize as if to state their basic postulate: *"Any of the physical properties of A which change with the application of heat may be observed and utilised for the measurement of temperature."* They do not themselves here use the term 'zeroth law of thermodynamics'. There are very many statements of these physical ideas in the physics literature long before this text, in very similar language. What was new here was just the label 'zeroth law of thermodynamics'. Fowler, with co-author Edward A. Guggenheim, wrote of the zeroth law as follows:

> ...we introduce the postulate: *If two assemblies are each in thermal equilibrium with a third assembly, they are in thermal equilibrium with each other.*

They then proposed that "it may be shown to follow that the condition for thermal equilibrium between several assemblies is the equality of a certain single-valued function of the thermodynamic states of the assemblies, which may be called the temperature t, any one of the assemblies being used as a "thermometer" reading the temperature t on a suitable scale. This postulate of the "*Existence of temperature*" could with advantage be known as *the zeroth law of thermodynamics*". The first sentence of this present article is a version of this statement. It is not explicitly evident in the existence statement of Fowler and Guggenheim that temperature refers to a unique attribute of a state of a system, such as is expressed in the idea of the hotness manifold. Also their statement refers explicitly to statistical mechanical assemblies, not explicitly to macroscopic thermodynamically defined systems.

First Law of Thermodynamics

The first law of thermodynamics is a version of the law of conservation of energy, adapted for thermodynamic systems. The law of conservation of energy states that the total energy of an isolated system is constant; energy can be transformed from one form to another, but cannot be created or destroyed. The first law is often formulated by stating that the change in the internal energy of a closed system is equal to the amount of heat supplied to the system, minus the amount of work done by the system on its surroundings. Equivalently, perpetual motion machines of the first kind are impossible.

History

Investigations into the nature of heat and work and their relationship began with the invention of the first engines used to extract water from mines. Improvements to such engines so as to increase their efficiency and power output came first from mechanics that tinkered with such machines but only slowly advanced the art. Deeper investigations that placed those on a mathematical and physics basis came later.

The process of development of the first law of thermodynamics was by way of much investigative trial and error over a period of about half a century. The first full statements of the law came in 1850 from Rudolf Clausius and from William Rankine; Rankine's statement was perhaps not quite as clear and distinct as was Clausius'. A main aspect of the struggle was to deal with the previously proposed caloric theory of heat.

Germain Hess in 1840 stated a conservation law for the so-called 'heat of reaction' for chemical reactions. His law was later recognized as a consequence of the first law of thermodynamics, but Hess's statement was not explicitly concerned with the relation between energy exchanges by heat and work.

According to Truesdell (1980), Julius Robert von Mayer in 1841 made a statement that meant that "in a process at constant pressure, the heat used to produce expansion is universally interconvertible with work", but this is not a general statement of the first law.

Original Statements: the "Thermodynamic Approach"

The original nineteenth century statements of the first law of thermodynamics appeared in a conceptual framework in which transfer of energy as heat was taken as a primitive notion, not defined or constructed by the theoretical development of the framework, but rather presupposed as prior to it and already accepted. The primitive notion of heat was taken as empirically established, especially through calorimetry regarded as a subject in its own right, prior to thermodynamics. Jointly primitive with this notion of heat were the notions of empirical temperature and thermal equilibrium. This framework also took as primitive the notion of transfer of energy as work. This framework did not presume a concept of energy in general, but regarded it as derived or synthesized from the prior notions of heat and work. By one author, this framework has been called the "thermodynamic" approach.

The first explicit statement of the first law of thermodynamics, by Rudolf Clausius in 1850, referred to cyclic thermodynamic processes.

> *In all cases in which work is produced by the agency of heat, a quantity of*

> *heat is consumed which is proportional to the work done; and conversely, by the expenditure of an equal quantity of work an equal quantity of heat is produced.*

Clausius also stated the law in another form, referring to the existence of a function of state of the system, the internal energy, and expressed it in terms of a differential equation for the increments of a thermodynamic process. This equation may described as follows:

> *In a thermodynamic process involving a closed system, the increment in the internal energy is equal to the difference between the heat accumulated by the system and the work done by it.*

Because of its definition in terms of increments, the value of the internal energy of a system is not uniquely defined. It is defined only up to an arbitrary additive constant of integration, which can be adjusted to give arbitrary reference zero levels. This non-uniqueness is in keeping with the abstract mathematical nature of the internal energy. The internal energy is customarily stated relative to a conventionally chosen standard reference state of the system.

The concept of internal energy is considered by Bailyn to be of "enormous interest". Its quantity cannot be immediately measured, but can only be inferred, by differencing actual immediate measurements. Bailyn likens it to the energy states of an atom, that were revealed by Bohr's energy relation $h\nu = E_{n''} - E_{n'}$. In each case, an unmeasurable quantity (the internal energy, the atomic energy level) is revealed by considering the difference of measured quantities (increments of internal energy, quantities of emitted or absorbed radiative energy).

Conceptual Revision: the "Mechanical Approach"

In 1907, George H. Bryan wrote about systems between which there is no transfer of matter (closed systems): "Definition. When energy flows from one system or part of a system to another otherwise than by the performance of mechanical work, the energy so transferred is called *heat*." This definition may be regarded as expressing a conceptual revision, as follows. This was systematically expounded in 1909 by Constantin Carathéodory, whose attention had been drawn to it by Max Born. Largely through Born's influence, this revised conceptual approach to the definition of heat came to be preferred by many twentieth-century writers. It might be called the "mechanical approach".

Energy can also be transferred from one thermodynamic system to another in association with transfer of matter. Born points out that in general such energy transfer is not resolvable uniquely into work and heat moieties. In general, when there is transfer of energy associated with matter transfer, work and heat transfers can be distinguished only when they pass through walls physically separate from those for matter transfer.

The "mechanical" approach postulates the law of conservation of energy. It also postulates that energy can be transferred from one thermodynamic system to another adiabatically as work, and that energy can be held as the internal energy of a thermodynamic system. It also postulates that energy can be transferred from one thermodynamic system to another by a path that is non-adiabatic, and is unaccompanied by matter transfer. Initially, it "cleverly" (according to Bailyn) refrains from labelling as 'heat' such non-adiabatic, unaccompanied transfer of energy. It rests on the primitive notion of *walls*, especially adiabatic walls and non-adiabatic walls, defined as follows. Temporarily, only for purpose of this definition, one can prohibit transfer of energy as work across a wall of interest. Then walls of interest fall into two classes, (a) those such that arbitrary systems separated by them remain independently in their own previously established respective states of internal thermodynamic equilibrium; they are defined as adiabatic; and (b) those without such independence; they are defined as non-adiabatic.

This approach derives the notions of transfer of energy as heat, and of temperature, as theoretical developments, not taking them as primitives. It regards calorimetry as a derived theory. It has an early origin in the nineteenth century, for example in the work of Helmholtz, but also in the work of many others.

Conceptually Revised Statement, According to the Mechanical Approach

The revised statement of the first law postulates that a change in the internal energy of a system due to any arbitrary process, that takes the system from a given initial thermodynamic state to a given final equilibrium thermodynamic state, can be determined through the physical existence, for those given states, of a reference process that occurs purely through stages of adiabatic work.

The revised statement is then

> *For a closed system, in any arbitrary process of interest that takes it from an initial to a final state of internal thermodynamic equilibrium, the change of internal energy is the same as that for a reference adiabatic work process that links those two states. This is so regardless of the path of the process of interest, and regardless of whether it is an adiabatic or a non-adiabatic process. The reference adiabatic work process may be chosen arbitrarily from amongst the class of all such processes.*

This statement is much less close to the empirical basis than are the original statements, but is often regarded as conceptually parsimonious in that it rests only on the concepts of adiabatic work and of non-adiabatic processes, not on the concepts of transfer of energy as heat and of empirical temperature that are presupposed by the original statements. Largely through the influence of Max Born, it is often regarded as theoretically preferable because of this conceptual parsimony. Born particularly observes that the revised approach avoids thinking in terms of what he calls the "imported engineering" concept of heat engines.

Basing his thinking on the mechanical approach, Born in 1921, and again in 1949, proposed to revise the definition of heat. In particular, he referred to the work of Constantin Carathéodory, who had in 1909 stated the first law without defining quantity of heat. Born's definition was specifically for transfers of energy without transfer of matter, and it has been widely followed in textbooks (examples:). Born observes that a transfer of matter between two systems is accompanied by a transfer of internal energy that cannot be resolved into heat and work components. There can be pathways to other systems, spatially separate from that of the matter transfer, that allow heat and work transfer independent of and simultaneous with the matter transfer. Energy is conserved in such transfers.

Description

The first law of thermodynamics for a closed system was expressed in two ways by Clausius. One way referred to cyclic processes and the inputs and outputs of the system, but did not refer to increments in the internal state of the system. The other way referred to an incremental change in the internal state of the system, and did not expect the process to be cyclic.

A cyclic process is one that can be repeated indefinitely often, returning the system to its initial state. Of particular interest for single cycle of a cyclic process are the net work done, and the net heat taken in (or 'consumed', in Clausius' statement), by the system.

In a cyclic process in which the system does net work on its surroundings, it is observed to be physically necessary not only that heat be taken into the system, but also, importantly, that some heat leave the system. The difference is the heat converted by the cycle into work. In each repetition of a cyclic process, the net work done by the system, measured in mechanical units, is proportional to the heat consumed, measured in calorimetric units.

The constant of proportionality is universal and independent of the system and in 1845 and 1847 was measured by James Joule, who described it as the *mechanical equivalent of heat*.

In a non-cyclic process, the change in the internal energy of a system is equal to net energy added as heat to the system minus the net work done by the system, both being measured in mechanical units. Taking ΔU as a change in internal energy, one writes

$$\Delta U = Q - W \quad \text{(sign convention of Clausius and generally in this article),}$$

where Q denotes the net quantity of heat supplied to the system by its surroundings and

W denotes the net work done by the system. This sign convention is implicit in Clausius' statement of the law given above. It originated with the study of heat engines that produce useful work by consumption of heat.

Often nowadays, however, writers use the IUPAC convention by which the first law is formulated with work done on the system by its surroundings having a positive sign. With this now often used sign convention for work, the first law for a closed system may be written:

$$\Delta U = Q + W \ \text{(signconventionofIUPAC)}.$$

This convention follows physicists such as Max Planck, and considers all net energy transfers to the system as positive and all net energy transfers from the system as negative, irrespective of any use for the system as an engine or other device.

When a system expands in a fictive quasistatic process, the work done by the system on the environment is the product, $P\,dV$, of pressure, P, and volume change, dV, whereas the work done *on* the system is $-P\,dV$. Using either sign convention for work, the change in internal energy of the system is:

$$dU = \delta Q - PdV \ \text{(quasi-static process)},$$

where δQ denotes the infinitesimal increment of heat supplied to the system from its surroundings.

Work and heat are expressions of actual physical processes of supply or removal of energy, while the internal energy U is a mathematical abstraction that keeps account of the exchanges of energy that befall the system. Thus the term heat for Q means "that amount of energy added or removed by conduction of heat or by thermal radiation", rather than referring to a form of energy within the system. Likewise, the term work energy for W means "that amount of energy gained or lost as the result of work". Internal energy is a property of the system whereas work done and heat supplied are not. A significant result of this distinction is that a given internal energy change ΔU can be achieved by, in principle, many combinations of heat and work.

Various Statements of the Law for Closed Systems

The law is of great importance and generality and is consequently thought of from several points of view. Most careful textbook statements of the law express it for closed systems. It is stated in several ways, sometimes even by the same author.

For the thermodynamics of closed systems, the distinction between transfers of energy as work and as heat is central and is within the scope of the present article. For the thermodynamics of open systems, such a distinction is beyond the scope of the present article, but some limited comments are made on it in the section below headed 'First law of thermodynamics for open systems'.

There are two main ways of stating a law of thermodynamics, physically or mathematically. They should be logically coherent and consistent with one another.

An example of a physical statement is that of Planck (1897/1903):

> It is in no way possible, either by mechanical, thermal, chemical, or other devices, to obtain perpetual motion, i.e. it is impossible to construct an engine which will work in a cycle and produce continuous work, or kinetic energy, from nothing.

This physical statement is restricted neither to closed systems nor to systems with states that are strictly defined only for thermodynamic equilibrium; it has meaning also for open systems and for systems with states that are not in thermodynamic equilibrium.

This statement by Crawford, for W, uses the sign convention of IUPAC, not that of Clausius. Though it does not explicitly say so, this statement refers to closed systems, and to internal energy U defined for bodies in states of thermodynamic equilibrium, which possess well-defined temperatures.

The history of statements of the law for closed systems has two main periods, before and after the work of Bryan (1907), of Carathéodory (1909), and the approval of Carathéodory's work given by Born (1921). The earlier traditional versions of the law for closed systems are nowadays often considered to be out of date.

Carathéodory's celebrated presentation of equilibrium thermodynamics refers to closed systems, which are allowed to contain several phases connected by internal walls of various kinds of impermeability and permeability (explicitly including walls that are permeable only to heat). Carathéodory's 1909 version of the first law of thermodynamics was stated in an axiom which refrained from defining or mentioning temperature or quantity of heat transferred. That axiom stated that the internal energy of a phase in equilibrium is a function of state, that the sum of the internal energies of the phases is the total internal energy of the system, and that the value of the total internal energy of the system is changed by the amount of work done adiabatically on it, considering work as a form of energy. That article considered this statement to be an expression of the law of conservation of energy for such systems. This version is nowadays widely accepted as authoritative, but is stated in slightly varied ways by different authors.

Such statements of the first law for closed systems assert the existence of internal energy as a function of state defined in terms of adiabatic work. Thus heat is not defined calorimetrically or as due to temperature difference. It is defined as a residual difference between change of internal energy and work done on the system, when that work does not account for the whole of the change of internal energy and the system is not adiabatically isolated.

The 1909 Carathéodory statement of the law in axiomatic form does not mention heat or temperature, but the equilibrium states to which it refers are explicitly defined by variable sets that necessarily include "non-deformation variables", such as pressures, which, within reasonable restrictions, can be rightly interpreted as empirical temperatures, and the walls connecting the phases of the system are explicitly defined as possibly impermeable to heat or permeable only to heat.

According to Münster (1970), "A somewhat unsatisfactory aspect of Carathéodory's theory is that a consequence of the Second Law must be considered at this point [in the statement of the first law], i.e. that it is not always possible to reach any state 2 from any other state 1 by means of an adiabatic process." Münster instances that no adiabatic process can reduce the internal energy of a system at constant volume. Carathéodory's paper asserts that its statement of the first law corresponds exactly to Joule's experimental arrangement, regarded as an instance of adiabatic work. It does not point out that Joule's experimental arrangement performed essentially irreversible work, through friction of paddles in a liquid, or passage of electric current through a resistance inside the system, driven by motion of a coil and inductive heating, or by an external current source, which can access the system only by the passage of electrons, and so is not strictly adiabatic, because electrons are a form of matter, which cannot penetrate adiabatic walls. The paper goes on to base its main argument on the possibility of quasi-static adiabatic work, which is essentially reversible. The paper asserts that it will avoid reference to Carnot cycles, and then proceeds to base its argument on cycles of forward and backward quasi-static adiabatic stages, with isothermal stages of zero magnitude.

Sometimes the concept of internal energy is not made explicit in the statement.

Sometimes the existence of the internal energy is made explicit but work is not explicitly mentioned in the statement of the first postulate of thermodynamics. Heat supplied is then defined as the residual change in internal energy after work has been taken into account, in a non-adiabatic process.

A respected modern author states the first law of thermodynamics as "Heat is a form of energy", which explicitly mentions neither internal energy nor adiabatic work. Heat is defined as energy transferred by thermal contact with a reservoir, which has a temperature, and is generally so large that addition and removal of heat do not alter its temperature. A current student text on chemistry defines heat thus: "*heat* is the exchange of thermal energy between a system and its surroundings caused by a temperature difference." The author then explains how heat is defined or measured by calorimetry, in terms of heat capacity, specific heat capacity, molar heat capacity, and temperature.

A respected text disregards the Carathéodory's exclusion of mention of heat from the statement of the first law for closed systems, and admits heat calorimetrically defined along with work and internal energy. Another respected text defines heat exchange as determined by temperature difference, but also mentions that the Born (1921) version is "completely rigorous". These versions follow the traditional approach that is now considered out of date, exemplified by that of Planck (1897/1903).

Evidence for the First Law of Thermodynamics for Closed Systems

The first law of thermodynamics for closed systems was originally induced from empirically observed evidence, including calorimetric evidence. It is nowadays, however, taken to provide the definition of heat via the law of conservation of energy and the definition of work in terms of changes in the external parameters of a system. The original discovery of the law was gradual over a period of perhaps half a century or more, and some early studies were in terms of cyclic processes.

The following is an account in terms of changes of state of a closed system through compound processes that are not necessarily cyclic. This account first considers processes for which the first law is easily verified because of their simplicity, namely adiabatic processes (in which there is no transfer as heat) and adynamic processes (in which there is no transfer as work).

Adiabatic Processes

In an adiabatic process, there is transfer of energy as work but not as heat. For all adiabatic process that takes a system from a given initial state to a given final state, irrespective of how the work is done, the respective eventual total quantities of energy transferred as work are one and the same, determined just by the given initial and final states. The work done on the system is defined and measured by changes in mechanical or quasi-mechanical variables external to the system. Physically, adiabatic transfer of energy as work requires the existence of adiabatic enclosures.

For instance, in Joule's experiment, the initial system is a tank of water with a paddle wheel inside. If we isolate the tank thermally, and move the paddle wheel with a pulley and a weight, we can relate the increase in temperature with the distance descended by the mass. Next, the system is returned to its initial state, isolated again, and the same amount of work is done on the tank using different devices (an electric motor, a chemical battery, a spring,...). In every case, the amount of work can be measured independently. The return to the initial state is not conducted by doing adiabatic work on the system. The evidence shows that the final state of the water (in particular, its temperature and volume) is the same in every case. It is irrelevant if the work is electrical, mechanical, chemical,... or if done suddenly or slowly, as long as it is performed in an adiabatic way, that is to say, without heat transfer into or out of the system.

Evidence of this kind shows that to increase the temperature of the water in the tank, the qualitative kind of adiabatically performed work does not matter. No qualitative kind of adiabatic work has ever been observed to decrease the temperature of the water in the tank.

A change from one state to another, for example an increase of both temperature and volume, may be conducted in several stages, for example by externally supplied electrical work on a resistor in the body, and adiabatic expansion allowing the body to do work on the surroundings. It needs to be shown that the time order of the stages, and their relative magnitudes, does not affect the amount of adiabatic work that needs to be done for the change of state. According to one respected scholar: "Unfortunately, it does not seem that experiments of this kind have ever been carried out carefully. ... We must therefore admit that the statement which we have enunciated here, and which is equivalent to the first law of thermodynamics, is not well founded on direct experimental evidence." Another expression of this view is "... no systematic precise experiments to verify this generalization directly have ever been attempted."

This kind of evidence, of independence of sequence of stages, combined with the above-mentioned evidence, of independence of qualitative kind of work, would show the existence of an important state variable that corresponds with adiabatic work, but not that such a state variable represented a conserved quantity. For the latter, another step of evidence is needed, which may be related to the concept of reversibility, as mentioned below.

That important state variable was first recognized and denoted U by Clausius in 1850, but he did not then name it, and he defined it in terms not only of work but also of heat transfer in the same process. It was also independently recognized in 1850 by Rankine, who also denoted it U ; and in 1851 by Kelvin who then called it «mechanical energy», and later «intrinsic energy». In 1865, after some hestitation, Clausius began calling his state function U "energy". In 1882 it was named as the *internal energy* by Helmholtz. If only adiabatic processes were of interest, and heat could be ignored, the concept of internal energy would hardly arise or be needed. The relevant physics would be largely covered by the concept of potential energy, as was intended in the 1847 paper of Helmholtz on the principle of conservation of energy, though that did not deal with forces that cannot be described by a potential, and thus did not fully justify the principle. Moreover, that paper was critical of the early work of Joule that had by then been performed. A great merit of the internal energy concept is that it frees thermodynamics from a restriction to cyclic processes, and allows a treatment in terms of thermodynamic states.

The fact of such irreversibility may be dealt with in two main ways, according to different points of view:

- Since the work of Bryan (1907), the most accepted way to deal with it nowadays, followed by Carathéodory, is to rely on the previously established concept of quasi-static processes, as follows. Actual physical processes of transfer of energy as work are always at least to some degree irreversible. The irreversibility is often due to mechanisms known as dissipative, that transform bulk kinetic energy into internal energy. Examples are friction and viscosity. If the process is performed more slowly, the frictional or viscous dissipation is less. In the limit of infinitely slow performance, the dissipation tends to zero and then the limiting process, though fictional rather than actual, is notionally reversible, and is called quasi-static. Throughout the course of the fictional limiting quasi-static process, the internal intensive variables of the system are equal to the external intensive variables, those that describe the reactive forces exerted by the surroundings. This can be taken to justify the formula

(1) $\quad W_{A \to O}^{\text{adiabatic, quasi-static}} = -W_{O \to A}^{\text{adiabatic, quasi-static}}$.

- Another way to deal with it is to allow that experiments with processes of heat transfer to or from the system may be used to justify the formula (1) above. Moreover, it deals to some extent with the problem of lack of direct experimental evidence that the time order of stages of a process does not matter in the determination of internal energy. This way does not provide theoretical purity in terms of adiabatic work processes, but is empirically feasible, and is in accord with experiments actually done, such as the Joule experiments mentioned just above, and with older traditions.

Adynamic Processes

A complementary observable aspect of the first law is about heat transfer. Adynamic transfer of energy as heat can be measured empirically by changes in the surroundings of the system of interest by calorimetry. This again requires the existence of adiabatic enclosure of the entire process, system and surroundings, though the separating wall between the surroundings and the system is thermally conductive or radiatively permeable, not adiabatic. A calorimeter can rely on measurement of sensible heat, which requires the existence of thermometers and measurement of temperature change in bodies of known sensible heat capacity under specified conditions; or it can rely on the measurement of latent heat, through measurement of masses of material that change phase, at temperatures fixed by the occurrence of phase changes under specified conditions in bodies of known latent heat of phase change. The calorimeter can be calibrated by adiabatically doing externally determined work on it. The most accurate method is by passing an electric ciurrent from outside through a resistance inside the calorimeter. The calibration allows comparison of calorimetric measurement of quantity of heat transferred with quantity of energy transferred as work. According to one textbook, "The most common device for measuring is an adiabatic bomb calorimeter." According to another textbook, "Calorimetry is widely used in present day laboratories." According to one opinion, "Most thermodynamic data come from calorimetry..." According to another opinion, "The most common method of measuring "heat" is with a calorimeter."

Overview of the Weight of Evidence for the Law

The first law of thermodynamics is so general that its predictions cannot all be directly tested. In many properly conducted experiments it has been precisely supported, and never violated. Indeed, within its scope of applicability, the law is so reliably established, that, nowadays, rather than experiment being considered as testing the accuracy of the law, it is more practical and realistic to think of the law as testing the accuracy of experiment. An experimental result that seems to violate the law may be assumed to be inaccurate or wrongly conceived, for example due to failure to account for an important physical factor. Thus, some may regard it as a principle more abstract than a law.

State Functional Formulation for Infinitesimal Processes

When the heat and work transfers in the equations above are infinitesimal in magnitude, they are often denoted by δ, rather than exact differentials denoted by d, as a reminder that heat and work do not describe the *state* of any system. The integral of an inexact differential depends upon the particular path taken through the space of thermodynamic parameters while the integral of an exact differential depends only upon the initial and final states. If the initial and final states are the same, then the integral of an inexact differential may or may not be zero, but the integral of an exact differential is always zero. The path taken by a thermodynamic system through a chemical or physical change is known as a thermodynamic process.

The first law for a closed homogeneous system may be stated in terms that include concepts that are established in the second law. The internal energy U may then be expressed as a function of the system's defining state variables S, entropy, and V, volume: $U = U(S, V)$. In these terms, T, the system's temperature, and P, its pressure, are partial derivatives of U with respect to S and V. These variables are important throughout thermodynamics, though not necessary for the statement of the first law. Rigorously, they are defined only when the system is in its own state of internal thermodynamic equilibrium. For some purposes, the concepts provide good approximations for scenarios sufficiently near to the system's internal thermodynamic equilibrium.

The first law requires that:

$$dU = \delta Q - \delta W \qquad \text{(closed system, general process, quasi-static or irreversible).}$$

Then, for the fictive case of a reversible process, dU can be written in terms of exact differentials. One may imagine reversible changes, such that there is at each instant negligible departure from thermodynamic equilibrium within the system. This excludes isochoric work. Then, mechanical work is given by $\delta W = -P\,dV$ and the quantity of heat added can be expressed as $\delta Q = T\,dS$. For these conditions

$$dU = Tds - PdV \qquad \text{(closed system, reversible process).}$$

While this has been shown here for reversible changes, it is valid in general, as U can be considered as a thermodynamic state function of the defining state variables S and V:

$$(2) \quad dU = TdS - PdV \quad \text{(closed system, general process, quasi-static or irreversible).}$$

Equation (2) is known as the fundamental thermodynamic relation for a closed system in the energy representation, for which the defining state variables are S and V, with respect to which T and P are partial derivatives of U. It is only in the fictive reversible case, when isochoric work is excluded, that the work done and heat transferred are given by $-P\,dV$ and $T\,dS$.

In the case of a closed system in which the particles of the system are of different types and, because chemical reactions may occur, their respective numbers are not necessarily constant, the fundamental thermodynamic relation for dU becomes:

$$dU = TdS - PdV + \sum_i \mu_i dN_i.$$

where dN_i is the (small) increase in amount of type-i particles in the reaction, and μ_i is known as the chemical potential of the type-i particles in the system. If dN_i is expressed in mol then μ_i is expressed in J/mol. If the system has more external mechanical variables than just the volume that can change, the fundamental thermodynamic relation further generalizes to:

$$dU = TdS - \sum_i X_i dx_i + \sum_j \mu_j dN_j.$$

Here the X_i are the generalized forces corresponding to the external variables x_i. The parameters X_i are independent of the size of the system and are called intensive parameters and the x_i are proportional to the size and called extensive parameters.

For an open system, there can be transfers of particles as well as energy into or out of the system during a process. For this case, the first law of thermodynamics still holds, in the form that the internal energy is a function of state and the change of internal energy in a process is a function only of its initial and final states, as noted in the section below headed First law of thermodynamics for open systems.

A useful idea from mechanics is that the energy gained by a particle is equal to the force applied to the particle multiplied by the displacement of the particle while that force is applied. Now consider the first law without the heating term: $dU = -PdV$. The pressure P can be viewed as a force (and in fact has units of force per unit area) while dV is the displacement (with units of distance times area). We may say, with respect to this work term, that a pressure difference forces a transfer of volume, and that the product of the two (work) is the amount of energy transferred out of the system as a result of the process. If one were to make this term negative then this would be the work done on the system.

It is useful to view the TdS term in the same light: here the temperature is known as a "generalized" force (rather than an actual mechanical force) and the entropy is a generalized displacement.

Similarly, a difference in chemical potential between groups of particles in the system drives a chemical reaction that changes the numbers of particles, and the corresponding product is the amount of chemical potential energy transformed in process. For example, consider a system consisting of two phases: liquid water and water vapor. There is a generalized "force" of evaporation that drives water molecules out of the liquid. There is a generalized "force" of condensation that drives vapor molecules out of the vapor. Only when these two "forces" (or chemical potentials) are equal is there equilibrium, and the net rate of transfer zero.

The two thermodynamic parameters that form a generalized force-displacement pair are called "conjugate variables". The two most familiar pairs are, of course, pressure-volume, and temperature-entropy.

Spatially Inhomogeneous Systems

Classical thermodynamics is initially focused on closed homogeneous systems (e.g. Planck 1897/1903), which might be regarded as 'zero-dimensional' in the sense that they have no spatial variation. But it is desired to study also systems with distinct internal motion and spatial inhomogeneity. For such systems, the principle of conservation of energy is expressed in terms not only of internal energy as defined for homogeneous systems, but also in terms of kinetic energy and potential energies of parts of the inhomogeneous system with respect to each other and with respect to long-range external forces. How the total energy of a system is allocated between these three more specific kinds of energy varies according to the purposes of different writers; this is because these components of energy are to some extent mathematical artefacts rather than actually measured physical quantities. For any closed homogeneous component of an inhomogeneous closed system, if E denotes the total energy of that component system, one may write

The distinction between internal and kinetic energy is hard to make in the presence of turbulent motion within the system, as friction gradually dissipates macroscopic kinetic energy of localised bulk flow into molecular random motion of molecules that is classified as internal energy. The rate of dissipation by friction of kinetic energy of localised bulk flow into internal energy, whether in turbulent or in streamlined flow, is an important quantity in non-equilibrium thermodynamics. This is a serious difficulty for attempts to define entropy for time-varying spatially inhomogeneous systems.

First Law of Thermodynamics for Open Systems

For the first law of thermodynamics, there is no trivial passage of physical conception from the closed system view to an open system view. For closed systems, the concepts of an adiabatic enclosure and of an adiabatic wall are fundamental. Matter and internal energy cannot permeate or penetrate such a wall. For an open system, there is a wall that allows penetration by matter. In general, matter in diffusive motion carries with it some internal energy, and some microscopic potential energy changes accompany the motion. An open system is not adiabatically enclosed.

There are some cases in which a process for an open system can, for particular purposes, be considered as if it were for a closed system. In an open system, by definition hypothetically or potentially, matter can pass between the system and its surroundings. But when, in a particular case, the process of interest involves only hypothetical or potential but no actual passage of matter, the process can be considered as if it were for a closed system.

Internal Energy for an Open System

Since the revised and more rigorous definition of the internal energy of a closed system rests upon the possibility of processes by which adiabatic work takes the system from one state to another, this leaves a problem for the definition of internal energy for an open system, for which adiabatic work is not in general possible. According to Max Born, the transfer of matter and energy across an open connection "cannot be reduced to mechanics". In contrast to the case of closed systems, for open systems, in the presence of diffusion, there is no unconstrained and unconditional physical distinction between convective transfer of internal energy by bulk flow of matter, the transfer of internal energy without transfer of matter (usually called heat conduction and work transfer), and change of various potential energies. The older traditional way and the conceptually revised (Carathéodory) way agree that there is no physically unique definition of heat and work transfer processes between open systems.

In particular, between two otherwise isolated open systems an adiabatic wall is by definition impossible. This problem is solved by recourse to the principle of conservation of energy. This principle allows a composite isolated system to be derived from two other component non-interacting isolated systems, in such a way that the total energy of the composite isolated system is equal to the sum of the total energies of the two component isolated systems. Two previously isolated systems can be subjected to the thermodynamic operation of placement between them of a wall permeable to matter and energy, followed by a time for establishment of a new thermodynamic state of internal equilibrium in the new single unpartitioned system. The internal energies of the initial two systems and of the final new system,

For the thermodynamic operation of adding two systems with internal energies U_1 and U_2, to produce a new system with internal energy U, one may write $U = U_1 + U_2$; the reference states for U, U_1 and U_2 should be specified accordingly, maintaining also that the internal energy of a system be proportional to its mass, so that the internal energies are extensive variables.

There is a sense in which this kind of additivity expresses a fundamental postulate that goes beyond the simplest ideas of classical closed system thermodynamics; the extensivity of some variables is not obvious, and needs explicit expression; indeed one author goes so far as to say that it could be recognized as a fourth law of thermodynamics, though this is not repeated by other authors.

Process of Transfer of Matter Between an Open System and Its Surroundings

A system connected to its surroundings only through contact by a single permeable wall, but otherwise isolated, is an open system. If it is initially in a state of contact equilibrium with a surrounding subsystem, a thermodynamic process of transfer of matter can be made to occur between them if the surrounding subsystem is subjected to some thermodynamic operation, for example, removal of a partition between it and some further surrounding subsystem. The removal of the partition in the surroundings initiates a process of exchange between the system and its contiguous surrounding subsystem.

An example is evaporation. One may consider an open system consisting of a collection of liquid, enclosed except where it is allowed to evaporate into or to receive condensate from its vapor above it, which may be considered as its contiguous surrounding subsystem, and subject to control of its volume and temperature.

A thermodynamic process might be initiated by a thermodynamic operation in the surroundings, that mechanically increases in the controlled volume of the vapor. Some mechanical work will be done within the surroundings by the vapor, but also some of the parent liquid will evaporate and enter the vapor collection which is the contiguous surrounding subsystem. Some internal energy will accompany the vapor that leaves the system, but it will not make sense to try to uniquely identify part of that internal energy as heat and part of it as work. Consequently, the energy transfer that accompanies the transfer of matter between the system and its surrounding subsystem cannot be uniquely split into heat and work transfers to or from the open system. The component of total energy transfer that accompanies the transfer of vapor into the surrounding subsystem is customarily called 'latent heat of evaporation', but this use of the word heat is a quirk of customary historical language, not in strict compliance with the thermodynamic definition of transfer of energy as heat. In this example, kinetic energy of bulk flow and potential energy with respect to long-range external forces such as gravity are both considered to be zero. The first law of thermodynamics refers to the change of internal energy of the open system, between its initial and final states of internal equilibrium.

Open System with Multiple Contacts

An open system can be in contact equilibrium with several other systems at once.

This includes cases in which there is contact equilibrium between the system, and several subsystems in its surroundings, including separate connections with subsystems through walls that are permeable to the transfer of matter and internal energy as heat and allowing friction of passage of the transferred matter, but immovable, and separate connections through adiabatic walls with others, and separate connections through diathermic walls impermeable to matter with yet others. Because there are physically separate connections that are permeable to energy but impermeable to matter, between the system and its surroundings, energy transfers between them can occur with definite heat and work characters. Conceptually essential here is that the internal energy transferred with the transfer of matter is measured by a variable that is mathematically independent of the variables that measure heat and work.

With such independence of variables, the total increase of internal energy in the process is then determined as the sum of the internal energy transferred from the surroundings with the transfer of matter through the walls that are permeable to it, and of the internal energy transferred to the system as heat through the diathermic walls, and of the energy transferred to the system as work through the adiabatic walls, including the energy transferred to the system by long-range forces. These simultaneously transferred quantities of energy are defined by events in the surroundings of the system. Because the internal energy transferred with matter is not in general uniquely resolvable into heat and work components, the total energy transfer cannot in general be uniquely resolved into heat and work components.

where there are n chemical constituents of the system and permeably connected surrounding subsystems, and where T, S, P, V, N_j, and μ_j, are defined as above.

For a general natural process, there is no immediate term-wise correspondence between equations (3) and (4), because they describe the process in different conceptual frames.

Nevertheless, a conditional correspondence exists. There are three relevant kinds of wall here: purely diathermal, adiabatic, and permeable to matter. If two of those kinds of wall are sealed off, leaving only one that permits transfers of energy, as work, as heat, or with matter, then the remaining permitted terms correspond precisely. If two of the kinds of wall are left unsealed, then energy transfer can be shared between them, so that the two remaining permitted terms do not correspond precisely.

Non-equilibrium Transfers

The transfer of energy between an open system and a single contiguous subsystem of its surroundings is considered also in non-equilibrium thermodynamics. The problem of definition arises also in this case. It may be allowed that the wall between the system and the subsystem is not only permeable to matter and to internal energy, but also may be movable so as to allow work to be done when the two systems have different pressures. In this case, the transfer of energy as heat is not defined.

Methods for study of non-equilibrium processes mostly deal with spatially continuous flow systems. In this case, the open connection between system and surroundings is usually taken to fully surround the system, so that there are no separate connections impermeable to matter but permeable to heat. Except for the special case mentioned above when there is no actual transfer of matter, which can be treated as if for a closed system, in strictly defined thermodynamic terms, it follows that transfer of energy as heat is not defined. In this sense, there is no such thing as 'heat flow' for a continuous-flow open system. Properly, for closed systems, one speaks of transfer of internal energy as heat, but in general, for open systems, one can speak safely only of transfer of internal energy. A factor here is that there are often cross-effects between distinct transfers, for example that transfer of one substance may cause transfer of another even when the latter has zero chemical potential gradient.

Usually transfer between a system and its surroundings applies to transfer of a state variable, and obeys a balance law, that the amount lost by the donor system is equal to the amount gained by the receptor system. Heat is not a state variable. For his 1947 definition of "heat transfer" for discrete open systems, the author Prigogine carefully explains at some length that his definition of it does not obey a balance law. He describes this as paradoxical.

The situation is clarified by Gyarmati, who shows that his definition of "heat transfer", for continuous-flow systems, really refers not specifically to heat, but rather to transfer of internal energy, as follows. He considers a conceptual small cell in a situation of continuous-flow as a system defined in the so-called Lagrangian way, moving with the local center of mass. The flow of matter across the boundary is zero when considered as a flow of total mass. Nevertheless, if the material constitution is of several chemically distinct components that can diffuse with respect to one another, the system is considered to be open, the diffusive flows of the components being defined with respect to the center of mass of the system, and balancing one another as to mass transfer. Still there can be a distinction between bulk flow of internal energy and diffusive flow of internal energy in this case, because the internal energy density does not have to be constant per unit mass of material, and allowing for non-conservation of internal energy because of local conversion of kinetic energy of bulk flow to internal energy by viscosity.

Gyarmati shows that his definition of "the heat flow vector" is strictly speaking a definition of flow of internal energy, not specifically of heat, and so it turns out that his use here of the word heat is contrary to the strict thermodynamic definition of heat, though it is more or less compatible with historical custom, that often enough did not clearly distinguish between heat and internal energy; he writes "that this relation must be considered to be the exact definition of the concept of heat flow, fairly loosely used in experimental physics and heat technics." Apparently in a different frame of thinking from that of the above-mentioned paradoxical usage in the earlier sections of the historic 1947 work by Prigogine, about discrete systems, this usage of Gyarmati is consistent with the later sections of the same 1947 work by Prigogine, about continuous-flow systems, which use the term "heat flux" in just this way. This usage is also followed by Glansdorff and Prigogine in their 1971 text about continuous-flow systems. They write: "Again the flow of internal energy may be split into a convection flow $\rho u \mathbf{v}$ and a conduction flow. This conduction flow is by definition the heat flow $\mathbf{W}$. Therefore: $\mathbf{j}[U] = \rho u \mathbf{v} + \mathbf{W}$ where u denotes the [internal] energy per unit mass. [These authors actually use the symbols E and e to denote internal energy but their notation has been changed here to accord with the notation of the present article. These authors actually use the symbol U to refer to total energy, including kinetic energy of bulk flow.]" This usage is followed also by other writers on non-equilibrium thermodynamics such as Lebon, Jou, and Casas-Vásquez, and de Groot and Mazur. This usage is described by Bailyn as stating the non-convective flow of internal energy, and is listed as his definition number 1, according to the first law of thermodynamics. This usage is also followed by workers in the kinetic theory of gases. This is not the *ad hoc* definition of "reduced heat flux" of Haase.

In the case of a flowing system of only one chemical constituent, in the Lagrangian representation, there is no distinction between bulk flow and diffusion of matter. Moreover, the flow of matter is zero into or out of the cell that moves with the local center of mass. In effect, in this description, one is dealing with a system effectively closed to the transfer of matter. But still one can validly talk of a distinction between bulk flow and diffusive flow of internal energy, the latter driven by a temperature gradient within the flowing material, and being defined with respect to the local center of mass of the bulk flow. In this case of a virtually closed system, because of the zero matter transfer, as noted above, one can safely distinguish between transfer of energy as work, and transfer of internal energy as heat.

Second Law of Thermodynamics

The second law of thermodynamics states that the total entropy of an isolated system always increases over time, or remains constant in ideal cases where the system is in a steady state or undergoing a reversible process. The increase in entropy accounts for the irreversibility of natural processes, and the asymmetry between future and past.

Historically, the second law was an empirical finding that was accepted as an axiom of thermodynamic theory. Statistical thermodynamics, classical or quantum, explains the microscopic origin of the law.

The second law has been expressed in many ways. Its first formulation is credited to the French scientist Sadi Carnot in 1824, who showed that there is an upper limit to the efficiency of conversion of heat to work in a heat engine.

Introduction

The first law of thermodynamics provides the basic definition of internal energy, associated with all thermodynamic systems, and states the rule of conservation of energy. The second law is concerned with the direction of natural processes. It asserts that a natural process runs only in one sense, and is not reversible. For example, heat always flows spontaneously from hotter to colder bodies, and never the reverse, unless external work is performed on the system. Its modern definition is in terms of entropy.

In a fictive reversible process, an infinitesimal increment in the entropy (dS) of a system is defined to result from an infinitesimal transfer of heat (δQ) to a closed system divided by the common temperature (T) of the system and the surroundings which supply the heat:

$$dS = \frac{\delta Q}{T} \qquad \text{(closed system, idealized fictive reversible process).}$$

Different notations are used for infinitesimal amounts of heat (δ) and infinitesimal amounts of entropy (d) because entropy is a function of state, while heat, like work, is not. For an actually possible infinitesimal process without exchange of matter with the surroundings, the second law requires that the increment in system entropy be greater than that:

$$dS > \frac{\delta Q}{T} \qquad \text{(closed system, actually possible, irreversible process).}$$

This is because a general process for this case may include work being done on the system by its surroundings, which must have frictional or viscous effects inside the system, and because heat transfer actually occurs only irreversibly, driven by a finite temperature difference.

The zeroth law of thermodynamics in its usual short statement allows recognition that two bodies in a relation of thermal equilibrium have the same temperature, especially that a test body has the same temperature as a reference thermometric body. For a body in thermal equilibrium with another, there are indefinitely many empirical temperature scales, in general respectively depending on the properties of a particular reference thermometric body. The second law allows a distinguished temperature scale, which defines an absolute, thermodynamic temperature, independent of the properties of any particular reference thermometric body.

Various Statements of the Law

The second law of thermodynamics may be expressed in many specific ways, the most prominent classical statements being the statement by Rudolf Clausius (1854), the statement by Lord Kelvin (1851), and the statement in axiomatic thermodynamics by Constantin Carathéodory (1909). These statements cast the law in general physical terms citing the impossibility of certain processes. The Clausius and the Kelvin statements have been shown to be equivalent.

Carnot's Principle

The historical origin of the second law of thermodynamics was in Carnot's principle. It refers to a cycle of a Carnot heat engine, fictively operated in the limiting mode of extreme slowness known as quasi-static, so that the heat and work transfers are between subsystems that are always in their own internal states of thermodynamic equilibrium. The Carnot engine is an idealized device of special interest to engineers who are concerned with the efficiency of heat engines. Carnot's principle was recognized by Carnot at a time when the caloric theory of heat was seriously considered, before the recognition of the first law of thermodynamics, and before the mathematical expression of the concept of entropy. Interpreted in the light of the first law, it is physically equivalent to the second law of thermodynamics, and remains valid today. It states

The efficiency of a quasi-static or reversible Carnot cycle depends only on the temperatures of the two heat reservoirs, and is the same, whatever the working substance. A Carnot engine operated in this way is the most efficient possible heat engine using those two temperatures.

Clausius Statement

The German scientist Rudolf Clausius laid the foundation for the second law of thermodynamics in 1850 by examining the relation between heat transfer and work. His formulation of the second law, which was published in German in 1854, is known as the *Clausius statement*:

Heat can never pass from a colder to a warmer body without some other change, connected therewith, occurring at the same time.

The statement by Clausius uses the concept of 'passage of heat'. As is usual in thermodynamic discussions, this means 'net transfer of energy as heat', and does not refer to contributory transfers one way and the other.

Heat cannot spontaneously flow from cold regions to hot regions without external work being performed on the system, which is evident from ordinary experience of refrigeration, for example. In a refrigerator, heat flows from cold to hot, but only when forced by an external agent, the refrigeration system.

Kelvin Statement

Lord Kelvin expressed the second law as

It is impossible, by means of inanimate material agency, to derive mechanical effect from any portion of matter by cooling it below the temperature of the coldest of the surrounding objects.

Equivalence of the Clausius and the Kelvin Statements

Suppose there is an engine violating the Kelvin statement: i.e., one that drains heat and converts it completely into work in a cyclic fashion without any other result. Now pair it with a reversed Carnot engine as shown by the figure. The net and sole effect of this newly created engine consisting of the two engines mentioned is transferring heat $\Delta Q = Q\left(\frac{1}{\eta} - 1\right)$ from the cooler reservoir to the hotter one, which violates the Clausius statement. Thus a violation of the Kelvin statement implies a violation of the Clausius statement, i.e. the Clausius statement implies the Kelvin statement. We can prove in a similar manner that the Kelvin statement implies the Clausius statement, and hence the two are equivalent.

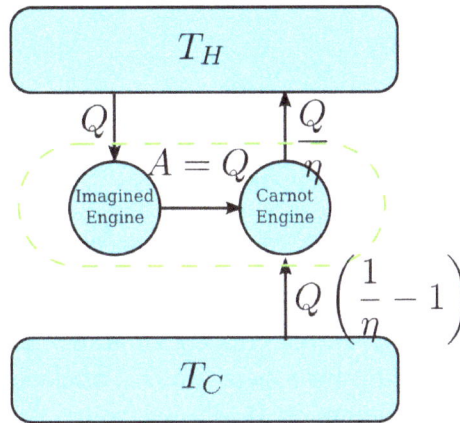

Derive Kelvin Statement from Clausius Statement

Planck's Proposition

Planck offered the following proposition as derived directly from experience. This is sometimes regarded as his statement of the second law, but he regarded it as a starting point for the derivation of the second law.

> *It is impossible to construct an engine which will work in a complete cycle, and produce no effect except the raising of a weight and cooling of a heat reservoir.*

Relation Between Kelvin's Statement and Planck's Proposition

It is almost customary in textbooks to speak of the "Kelvin-Planck statement" of the law, as for example in the text by ter Haar and Wergeland. One text gives a statement very like Planck's proposition, but attributes it to Kelvin without mention of Planck. One monograph quotes Planck's proposition as the "Kelvin-Planck" formulation, the text naming Kelvin as its author, though it correctly cites Planck in its references. The reader may compare the two statements quoted just above here.

Planck's Statement

Planck stated the second law as follows.

> *Every process occurring in nature proceeds in the sense in which the sum of the entropies of all bodies taking part in the process is increased. In the limit, i.e. for reversible processes, the sum of the entropies remains unchanged.*

Rather like Planck's statement is that of Uhlenbeck and Ford for *irreversible phenomena.*

... in an irreversible or spontaneous change from one equilibrium state to another (as for example the equalization of temperature of two bodies A and B, when brought in contact) the entropy always increases.

Principle of Carathéodory

Constantin Carathéodory formulated thermodynamics on a purely mathematical axiomatic foundation. His statement of the second law is known as the Principle of Carathéodory, which may be formulated as follows:

In every neighborhood of any state S of an adiabatically enclosed system there are states inaccessible from S.

With this formulation, he described the concept of adiabatic accessibility for the first time and provided the foundation for a new subfield of classical thermodynamics, often called geometrical thermodynamics. It follows from Carathéodory's principle that quantity of energy quasi-statically transferred as heat is a holonomic process function, in other words, $\delta Q = TdS$.

Though it is almost customary in textbooks to say that Carathéodory's principle expresses the second law and to treat it as equivalent to the Clausius or to the Kelvin-Planck statements, such is not the case. To get all the content of the second law, Carathéodory's principle needs to be supplemented by Planck's principle, that isochoric work always increases the internal energy of a closed system that was initially in its own internal thermodynamic equilibrium.

Planck's Principle

In 1926, Max Planck wrote an important paper on the basics of thermodynamics. He indicated the principle

The internal energy of a closed system is increased by an adiabatic process, throughout the duration of which, the volume of the system remains constant.

This formulation does not mention heat and does not mention temperature, nor even entropy, and does not necessarily implicitly rely on those concepts, but it implies the content of the second law. A closely related statement is that "Frictional pressure never does positive work." Using a now-obsolete form of words, Planck himself wrote: "The production of heat by friction is irreversible."

Not mentioning entropy, this principle of Planck is stated in physical terms. It is very closely related to the Kelvin statement given just above. It is relevant that for a system at constant volume and mole numbers, the entropy is a monotonic function of the internal energy. Nevertheless, this principle of Planck is not actually Planck's preferred statement of the second law, which is quoted above, in a previous sub-section of the present section of this present article, and relies on the concept of entropy.

A statement that in a sense is complementary to Planck's principle is made by Borgnakke and Sonntag. They do not offer it as a full statement of the second law:

> ... there is only one way in which the entropy of a [closed] system can be decreased, and that is to transfer heat from the system.

Differing from Planck's just foregoing principle, this one is explicitly in terms of entropy change. Of course, removal of matter from a system can also decrease its entropy.

Statement for a System that has a Known Expression of Its Internal Energy as a Function of Its Extensive State Variables

The second law has been shown to be equivalent to the internal energy U being a weakly convex function, when written as a function of extensive properties (mass, volume, entropy, ...).

Corollaries

Perpetual Motion of the Second Kind

Before the establishment of the Second Law, many people who were interested in inventing a perpetual motion machine had tried to circumvent the restrictions of first law of thermodynamics by extracting the massive internal energy of the environment as the power of the machine. Such a machine is called a "perpetual motion machine of the second kind". The second law declared the impossibility of such machines.

Carnot Theorem

Carnot's theorem (1824) is a principle that limits the maximum efficiency for any possible engine. The efficiency solely depends on the temperature difference between the hot and cold thermal reservoirs. Carnot's theorem states:

- All irreversible heat engines between two heat reservoirs are less efficient than a Carnot engine operating between the same reservoirs.

- All reversible heat engines between two heat reservoirs are equally efficient with a Carnot engine operating between the same reservoirs.

In his ideal model, the heat of caloric converted into work could be reinstated by reversing the motion of the cycle, a concept subsequently known as thermodynamic reversibility. Carnot, however, further postulated that some caloric is lost, not being converted to mechanical work. Hence, no real heat engine could realise the Carnot cycle's reversibility and was condemned to be less efficient.

Though formulated in terms of caloric, rather than entropy, this was an early insight into the second law.

Clausius Inequality

The Clausius theorem (1854) states that in a cyclic process

$$\oint \frac{\delta Q}{T} \leq 0.$$

The equality holds in the reversible case and the '<' is in the irreversible case. The reversible case is used to introduce the state function entropy. This is because in cyclic processes the variation of a state function is zero from state functionality.

Thermodynamic Temperature

For an arbitrary heat engine, the efficiency is:

$$\eta = \frac{W_n}{q_H} = \frac{q_H - q_C}{q_H} = 1 - \frac{q_C}{q_H} \tag{1}$$

where W_n is for the net work done per cycle. Thus the efficiency depends only on q_C/q_H.

Carnot's theorem states that all reversible engines operating between the same heat reservoirs are equally efficient. Thus, any reversible heat engine operating between temperatures T_1 and T_2 must have the same efficiency, that is to say, the efficiency is the function of temperatures only: $\dfrac{q_C}{q_H} = f(T_H, T_C) \tag{2}$

In addition, a reversible heat engine operating between temperatures T_1 and T_3 must have the same efficiency as one consisting of two cycles, one between T_1 and another (intermediate) temperature T_2, and the second between T_2 and T_3. This can only be the case if

$$f(T_1, T_3) = \frac{q_3}{q_1} = \frac{q_2 q_3}{q_1 q_2} = f(T_1, T_2) f(T_2, T_3).$$

Now consider the case where T_1 is a fixed reference temperature: the temperature of the triple point of water. Then for any T_2 and T_3,

$$f(T_2, T_3) = \frac{f(T_1, T_3)}{f(T_1, T_2)} = \frac{273.16 \cdot f(T_1, T_3)}{273.16 \cdot f(T_1, T_2)}.$$

Therefore, if thermodynamic temperature is defined by

$$T = 273.16 \cdot f(T_1, T)$$

then the function f, viewed as a function of thermodynamic temperature, is simply

$$f(T_2, T_3) = \frac{T_3}{T_2},$$

and the reference temperature T_1 will have the value 273.16. (Of course any reference temperature and any positive numerical value could be used—the choice here corresponds to the Kelvin scale.)

Entropy

According to the Clausius equality, for a reversible process

$$\oint \frac{\delta Q}{T} = 0$$

That means the line integral $\int_L \frac{\delta Q}{T}$ is path independent.

So we can define a state function S called entropy, which satisfies

$$dS = \frac{\delta Q}{T}$$

With this we can only obtain the difference of entropy by integrating the above formula. To obtain the absolute value, we need the Third Law of Thermodynamics, which states that S=0 at absolute zero for perfect crystals.

For any irreversible process, since entropy is a state function, we can always connect the initial and terminal states with an imaginary reversible process and integrating on that path to calculate the difference in entropy.

Now reverse the reversible process and combine it with the said irreversible process. Applying Clausius inequality on this loop,

$$-\Delta S + \int \frac{\delta Q}{T} = \oint \frac{\delta Q}{T} < 0$$

Thus,

$$\Delta S \geq \int \frac{\delta Q}{T}$$

where the equality holds if the transformation is reversible.

Notice that if the process is an adiabatic process, then $\delta Q = 0$, so $\Delta S \geq 0$.

Energy, Available Useful Work

An important and revealing idealized special case is to consider applying the Second Law to the scenario of an isolated system (called the total system or universe), made up of two parts: a sub-system of interest, and the sub-system's surroundings. These surroundings are imagined to be so large that they can be considered as an *unlimited* heat reservoir at temperature T_R and pressure P_R — so that no matter how much heat is transferred to (or from) the sub-system, the temperature of the surroundings will remain T_R; and no matter how much the volume of the sub-system expands (or contracts), the pressure of the surroundings will remain P_R.

Whatever changes to dS and dS_R occur in the entropies of the sub-system and the surroundings individually, according to the Second Law the entropy S_{tot} of the isolated total system must not decrease:

$$dS_{tot} = dS + dS_R \geq 0$$

According to the First Law of Thermodynamics, the change dU in the internal energy of the sub-system is the sum of the heat δq added to the sub-system, *less* any work δw done *by* the sub-system, *plus* any net chemical energy entering the sub-system $d \sum \mu_{iR} N_i$, so that:

$$dU = \delta q - \delta w + d\left(\sum \mu_{iR} N_i\right)$$

where μ_{iR} are the chemical potentials of chemical species in the external surroundings.

Now the heat leaving the reservoir and entering the sub-system is

$$\delta q = T_R(-dS_R) \leq T_R dS$$

where we have first used the definition of entropy in classical thermodynamics (alternatively, in statistical thermodynamics, the relation between entropy change, temperature and absorbed heat can be derived); and then the Second Law inequality from above.

It therefore follows that any net work δw done by the sub-system must obey

$$\delta w \leq -dU + T_R dS + \sum \mu_{iR} dN_i$$

It is useful to separate the work δw done by the subsystem into the *useful* work δw_u that can be done *by* the sub-system, over and beyond the work $p_R\, dV$ done merely by the sub-system expanding against the surrounding external pressure, giving the following relation for the useful work (exergy) that can be done:

$$\delta w_u \leq -d\left(U - T_R S + p_R V - \sum \mu_{iR} N_i\right)$$

It is convenient to define the right-hand-side as the exact derivative of a thermodynamic potential, called the *availability* or *exergy* E of the subsystem,

$$E = U - T_R S + p_R V - \sum \mu_{iR} N_i$$

The Second Law therefore implies that for any process which can be considered as divided simply into a subsystem, and an unlimited temperature and pressure reservoir with which it is in contact,

$$dE + \delta w_u \leq 0$$

i.e. the change in the subsystem's exergy plus the useful work done *by* the subsystem (or, the change in the subsystem's exergy less any work, additional to that done by the pressure reservoir, done *on* the system) must be less than or equal to zero.

In sum, if a proper *infinite-reservoir-like* reference state is chosen as the system surroundings in the real world, then the Second Law predicts a decrease in E for an irreversible process and no change for a reversible process.

$$dS_{tot} \geq 0 \text{ Is equivalent to } dE + \delta w_u \leq 0$$

This expression together with the associated reference state permits a design engineer working at the macroscopic scale (above the thermodynamic limit) to utilize the Second Law without directly measuring or considering entropy change in a total isolated system. Those changes have already been considered by the assumption that the system under consideration can reach equilibrium with the reference state without altering the reference state. An efficiency for a process or collection of processes that compares it to the reversible ideal may also be found.

This approach to the Second Law is widely utilized in engineering practice, environmental accounting, systems ecology, and other disciplines.

History

Nicolas Léonard Sadi Carnot in the traditional uniform of a student of the École Polytechnique.

The first theory of the conversion of heat into mechanical work is due to Nicolas Léonard Sadi Carnot in 1824. He was the first to realize correctly that the efficiency of this conversion depends on the difference of temperature between an engine and its environment.

Recognizing the significance of James Prescott Joule's work on the conservation of energy, Rudolf Clausius was the first to formulate the second law during 1850, in this form: heat does not flow *spontaneously* from cold to hot bodies. While common knowledge now, this was contrary to the caloric theory of heat popular at the time, which considered heat as a fluid. From there he was able to infer the principle of Sadi Carnot and the definition of entropy (1865).

Established during the 19th century, the Kelvin-Planck statement of the Second Law says, "It is impossible for any device that operates on a cycle to receive heat from a single reservoir and produce a net amount of work." This was shown to be equivalent to the statement of Clausius.

The ergodic hypothesis is also important for the Boltzmann approach. It says that, over long periods of time, the time spent in some region of the phase space of microstates with the same energy is proportional to the volume of this region, i.e. that all accessible microstates are equally probable over a long period of time. Equivalently, it says that time average and average over the statistical ensemble are the same.

There is a traditional doctrine, starting with Clausius, that entropy can be understood in terms of molecular 'disorder' within a macroscopic system. This doctrine is obsolescent.

Account Given by Clausius

Rudolf Clausius

In 1856, the German physicist Rudolf Clausius stated what he called the "second fundamental theorem in the mechanical theory of heat" in the following form:

$$\int \frac{\delta Q}{T} = -N$$

where Q is heat, T is temperature and N is the "equivalence-value" of all uncompensated transformations involved in a cyclical process. Later, in 1865, Clausius would come to define "equivalence-value" as entropy. On the heels of this definition, that same year, the most famous version of the second law was read in a presentation at the Philosophical Society of Zurich on April 24, in which, in the end of his presentation, Clausius concludes:

The entropy of the universe tends to a maximum.

This statement is the best-known phrasing of the second law. Because of the looseness of its language, e.g. universe, as well as lack of specific conditions, e.g. open, closed, or isolated, many people take this simple statement to mean that the second law of thermodynamics applies virtually to every subject imaginable. This, of course, is not true; this statement is only a simplified version of a more extended and precise description.

In terms of time variation, the mathematical statement of the second law for an isolated system undergoing an arbitrary transformation is:

$$\frac{dS}{dt} \geq 0$$

where

S is the entropy of the system and

t is time.

The equality sign applies after equilibration. An alternative way of formulating of the second law for isolated systems is:

$$\frac{dS}{dt} = \dot{S}_i \text{ with } \dot{S}_i \geq 0$$

with $\dot{S}_i$ the sum of the rate of entropy production by all processes inside the system. The advantage of this formulation is that it shows the effect of the entropy production. The rate of entropy production is a very important concept since it determines (limits) the efficiency of thermal machines. Multiplied with ambient temperature T_a it gives the so-called dissipated energy $P_{diss} = T_a \dot{S}_i$.

The expression of the second law for closed systems (so, allowing heat exchange and moving boundaries, but not exchange of matter) is:

$$\frac{dS}{dt} = \frac{\dot{Q}}{T} + \dot{S}_i \text{ with } \dot{S}_i \geq 0$$

Here

$\dot{Q}$ is the heat flow into the system

T is the temperature at the point where the heat enters the system.

The equality sign holds in the case that only reversible processes take place inside the system. If irreversible processes take place (which is the case in real systems in operation) the >-sign holds. If heat is supplied to the system at several places we have to take the algebraic sum of the corresponding terms.

For open systems (also allowing exchange of matter):

$$\frac{dS}{dt} = \frac{\dot{Q}}{T} + \dot{S} + \dot{S}_i \text{ with } \dot{S}_i \geq 0$$

Here $\dot{S}$ is the flow of entropy into the system associated with the flow of matter entering the system. It should not be confused with the time derivative of the entropy. If matter is supplied at several places we have to take the algebraic sum of these contributions.

Statistical Mechanics

Statistical mechanics gives an explanation for the second law by postulating that a material is composed of atoms and molecules which are in constant motion. A particular set of positions and velocities for each particle in the system is called a microstate of the system and because of the constant motion, the system is constantly changing its microstate. Statistical mechanics postulates that, in equilibrium, each microstate that the system might be in is equally likely to occur, and when this assumption is made, it leads directly to the conclusion that the second law must hold in a statistical sense. That is, the second law will hold on average, with a statistical variation on the order of $1/\sqrt{N}$ where N is the number of particles in the system. For everyday (macroscopic) situations, the probability that the second law will be violated is practically zero. However, for systems with a small number of particles, thermodynamic parameters, including the entropy, may show significant statistical deviations from that predicted by the second law. Classical thermodynamic theory does not deal with these statistical variations.

Derivation from Statistical Mechanics

Due to Loschmidt's paradox, derivations of the Second Law have to make an assumption regarding the past, namely that the system is uncorrelated at some time in the past; this allows for simple probabilistic treatment. This assumption is usually thought as a boundary condition, and thus the second Law is ultimately a consequence of the initial conditions somewhere in the past, probably at the beginning of the universe (the Big Bang), though other scenarios have also been suggested.

Given these assumptions, in statistical mechanics, the Second Law is not a postulate, rather it is a consequence of the fundamental postulate, also known as the equal prior probability postulate, so long as one is clear that simple probability arguments are applied only to the future, while for the past there are auxiliary sources of information which tell us that it was low entropy. The first part of the second law, which states that the entropy of a thermally isolated system can only increase, is a trivial consequence of the equal prior probability postulate, if we restrict the notion of the entropy to systems in thermal equilibrium. The entropy of an isolated system in thermal equilibrium containing an amount of energy of E is:

$$S = k_B \ln\left[\Omega(E)\right]$$

Derivation of the Entropy Change for Reversible Processes

The second part of the Second Law states that the entropy change of a system undergoing a reversible process is given by:

$$dS = \frac{\delta Q}{T}$$

where the temperature is defined as:

$$\frac{1}{k_{\mathrm{B}}T} \equiv \beta \equiv \frac{d \ln\left[\Omega(E)\right]}{dE}$$

Suppose that the system has some external parameter, x, that can be changed. In general, the energy eigenstates of the system will depend on x. According to the adiabatic theorem of quantum mechanics, in the limit of an infinitely slow change of the system's Hamiltonian, the system will stay in the same energy eigenstate and thus change its energy according to the change in energy of the energy eigenstate it is in.

The generalized force, X, corresponding to the external variable x is defined such that Xdx is the work performed by the system if x is increased by an amount dx. E.g., if x is the volume, then X is the pressure. The generalized force for a system known to be in energy eigenstate E_r is given by:

$$X = -\frac{dE_r}{dx}$$

Since the system can be in any energy eigenstate within an interval of δE, we define the generalized force for the system as the expectation value of the above expression:

$$X = -\left\langle \frac{dE_r}{dx} \right\rangle$$

Living Organisms

There are two principal ways of formulating thermodynamics, (a) through passages from one state of thermodynamic equilibrium to another, and (b) through cyclic processes, by which the system is left unchanged, while the total entropy of the surroundings is increased. These two ways help to understand the processes of life. This topic is mostly beyond the scope of this present article, but has been considered by several authors, such as Erwin Schrödinger, Léon Brillouin and Isaac Asimov. It is also the topic of current research.

To a fair approximation, living organisms may be considered as examples of (b). Approximately, an animal's physical state cycles by the day, leaving the animal nearly unchanged. Animals take in food, water, and oxygen, and, as a result of metabolism, give out breakdown products and heat. Plants take in radiative energy from the sun, which may be regarded as heat, and carbon dioxide and water. They give out oxygen. In this way they grow. Eventually they die, and their remains rot. This can be regarded as a cyclic process. Overall, the sunlight is from a high temperature source, the sun, and its energy is passed to a lower temperature sink, the soil. This is an increase of entropy of the surroundings of the plant. Thus animals and plants obey the second law of thermodynamics, considered in terms of cyclic processes. Simple concepts of efficiency of heat engines are hardly applicable to this problem because they assume closed systems.

From the thermodynamic viewpoint that considers (a), passages from one equilibrium state to another, only a roughly approximate picture appears, because living organisms are never in states of thermodynamic equilibrium. Living organisms must often be considered as open systems, because they take in nutrients and give out waste products. Thermodynamics of open systems is currently often considered in terms of passages from one state of thermodynamic equilibrium to another, or in terms of flows in the approximation of local thermodynamic equilibrium. The problem for living organisms may be further simplified by the approximation of assuming a steady state with unchanging flows. General principles of entropy production for such approximations are subject to unsettled current debate or research. Nevertheless, ideas derived from this viewpoint on the second law of thermodynamics are enlightening about living creatures.

Gravitational Systems

In systems that do not require for their descriptions the general theory of relativity, bodies always have positive heat capacity, meaning that the temperature rises with energy. Therefore, when energy flows from a high-temperature object to a low-temperature object, the source temperature is decreased while the sink temperature is increased; hence temperature differences tend to diminish over time. This is not always the case for systems in which the gravitational force is important and the general theory of relativity is required. Such systems can spontaneously change towards uneven spread of mass and energy. This applies to the universe in large scale, and consequently it may be difficult or impossible to apply the second law to it. Beyond this, the thermodynamics of systems described by the general theory of relativity is beyond the scope of the present article.

Non-equilibrium States

The theory of classical or equilibrium thermodynamics is idealized. A main postulate or assumption, often not even explicitly stated, is the existence of systems in their own internal states of thermodynamic equilibrium. In general, a region of space containing a physical system at a given time, that may be found in nature, is not in thermodynamic equilibrium, read in the most stringent terms. In looser terms, nothing in the entire universe is or has ever been truly in exact thermodynamic equilibrium.

For purposes of physical analysis, it is often enough convenient to make an assumption of thermodynamic equilibrium. Such an assumption may rely on trial and error for its justification. If the assumption is justified, it can often be very valuable and useful because it makes available the theory of thermodynamics. Elements of the equilibrium assumption are that a system is observed to be unchanging over an indefinitely long time, and that there are so many particles in a system, that its particulate nature can be entirely ignored. Under such an equilibrium assumption, in general, there are no macroscopically detectable fluctuations. There is an exception, the case of critical states, which exhibit to the naked eye the phenomenon of critical opalescence. For laboratory studies of critical states, exceptionally long observation times are needed.

In all cases, the assumption of thermodynamic equilibrium, once made, implies as a consequence that no putative candidate "fluctuation" alters the entropy of the system.

It can easily happen that a physical system exhibits internal macroscopic changes that are fast enough to invalidate the assumption of the constancy of the entropy. Or that a physical system has so few particles that the particulate nature is manifest in observable fluctuations. Then the assumption of thermodynamic equilibrium is to be abandoned. There is no unqualified general definition of entropy for non-equilibrium states.

There are intermediate cases, in which the assumption of local thermodynamic equilibrium is a very good approximation, but strictly speaking it is still an approximation, not theoretically ideal. For non-equilibrium situations in general, it may be useful to consider statistical mechanical definitions of other quantities that may be conveniently called 'entropy', but they should not be confused or conflated with thermodynamic entropy properly defined for the second law. These other quantities indeed belong to statistical mechanics, not to thermodynamics, the primary realm of the second law.

The physics of macroscopically observable fluctuations is beyond the scope of this article.

Arrow of Time

The second law of thermodynamics is a physical law that is not symmetric to reversal of the time direction.

The second law has been proposed to supply an explanation of the difference between moving forward and backwards in time, such as why the cause precedes the effect (the causal arrow of time).

Irreversibility

Irreversibility in thermodynamic processes is a consequence of the asymmetric character of thermodynamic operations, and not of any internally irreversible microscopic properties of the bodies. Thermodynamic operations are macroscopic external interventions imposed on the participating bodies, not derived from their internal properties. There are reputed "paradoxes" that arise from failure to recognize this.

Loschmidt's Paradox

Loschmidt's paradox, also known as the reversibility paradox, is the objection that it should not be possible to deduce an irreversible process from the time-symmetric dynamics that describe the microscopic evolution of a macroscopic system.

In the opinion of Schrödinger, "It is now quite obvious in what manner you have to reformulate the law of entropy—or for that matter, all other irreversible statements—so that they be capable of being derived from reversible models. You must not speak of one isolated system but at least of two, which you may for the moment consider isolated from the rest of the world, but not always from each other." The two systems are isolated from each other by the wall, until it is removed by the thermodynamic operation, as envisaged by the law. The thermodynamic operation is externally imposed, not subject to the reversible microscopic dynamical laws that govern the constituents of the systems. It is the cause of the irreversibility. The statement of the law in this present article complies with Schrödinger's advice. The cause–effect relation is logically prior to the second law, not derived from it.

Poincaré Recurrence Theorem

The Poincaré recurrence theorem considers a theoretical microscopic description of an isolated physical system. This may be considered as a model of a thermodynamic system after a thermodynamic operation has removed an internal wall. The system will, after a sufficiently long time, return to a microscopically defined state very close to the initial one. The Poincaré recurrence time is the length of time elapsed until the return. It is exceedingly long, likely longer than the life of the universe, and depends sensitively on the geometry of the wall that was removed by the thermodynamic operation. The recurrence theorem may be perceived as apparently contradicting the second law of thermodynamics. More obviously, however, it is simply a microscopic model of thermodynamic equilibrium in an isolated system formed by removal of a wall between two systems. For a typical thermodynamical system, the recurrence time is so large (many many times longer than the lifetime of the universe) that, for all practical purposes, one cannot observe the recurrence. One might wish, nevertheless, to imagine that one could wait for the Poincaré recurrence, and then re-insert the wall that was removed by the thermodynamic operation. It is then evident that the appearance of irreversibility is due to the utter unpredictability of the Poincaré recurrence given only that the initial state was one of thermodynamic equilibrium, as is the case in macroscopic thermodynamics. Even if one could wait for it, one has no practical possibility of picking the right instant at which to re-insert the wall. The Poincaré recurrence theorem provides a solution to Loschmidt's paradox. If an isolated thermodynamic system could be monitored over increasingly many multiples of the average Poincaré recurrence time, the thermodynamic behavior of the system would become invariant under time reversal.

Maxwell's Demon

James Clerk Maxwell imagined one container divided into two parts, A and B. Both parts are filled with the same gas at equal temperatures and placed next to each other, separated by a wall. Observing the molecules on both sides, an imaginary demon guards a microscopic trapdoor in the wall. When a faster-than-average molecule from A flies towards the trapdoor, the demon opens it, and the molecule will fly from A to B. The average speed of the molecules in B will have increased while in A they will have slowed down on average. Since average molecular speed corresponds to temperature, the temperature decreases in A and increases in B, contrary to the second law of thermodynamics.

One response to this question was suggested in 1929 by Leó Szilárd and later by Léon Brillouin. Szilárd pointed out that a real-life Maxwell's demon would need to have some means of measuring molecular speed, and that the act of acquiring information would require an expenditure of energy.

Maxwell's demon repeatedly alters the permeability of the wall between A and B. It is therefore performing thermodynamic operations on a microscopic scale, not just observing ordinary spontaneous or natural macroscopic thermodynamic processes.

Quotations

The law that entropy always increases holds, I think, the supreme position among the laws of Nature. If someone points out to you that your pet theory of the universe is in disagreement with Maxwell's equations — then so much the worse for Maxwell's equations. If it is found to be contradicted by observation — well, these experimentalists do bungle things sometimes. But if your theory is found to be against the second law of thermodynamics I can give you no hope; there is nothing for it but to collapse in deepest humiliation.

> — Sir Arthur Stanley Eddington, The Nature of the Physical World (1927)

There have been nearly as many formulations of the second law as there have been discussions of it.

> — Philosopher / Physicist P.W. Bridgman, (1941)

Clausius is the author of the sibyllic utterance, "The energy of the universe is constant; the entropy of the universe tends to a maximum." The objectives of continuum thermomechanics stop far short of explaining the "universe", but within that theory we may easily derive an explicit statement in some ways reminiscent of Clausius, but referring only to a modest object: an isolated body of finite size.

— *Truesdell, C., Muncaster, R.G. (1980). Fundamentals of Maxwell's Kinetic Theory of a Simple Monatomic Gas, Treated as a Branch of Rational Mechanics, Academic Press, New York, ISBN 0-12-701350-4, p.17.*

Third Law of Thermodynamics

The third law of thermodynamics is sometimes stated as follows, regarding the properties of systems in equilibrium at absolute zero temperature:

The entropy of a perfect crystal at absolute zero is exactly equal to zero.

At absolute zero (zero kelvin), the system must be in a state with the minimum possible energy, and the above statement of the third law holds true provided that the perfect crystal has only one minimum energy state. Entropy is related to the number of accessible microstates, and for a system consisting of many particles, quantum mechanics indicates that there is only one unique state (called the ground state) with minimum energy. If the system does not have a well-defined order (if its order is glassy, for example), then in practice there will remain some finite entropy as the system is brought to very low temperatures as the system becomes locked into a configuration with non-minimal energy. The constant value is called the residual entropy of the system.

The Nernst–Simon statement of the third law of thermodynamics concerns thermodynamic processes at a fixed, low temperature:

The entropy change associated with any condensed system undergoing a reversible isothermal process approaches zero as the temperature at which it is performed approaches 0 K.

Here a condensed system refers to liquids and solids. A classical formulation by Nernst (actually a consequence of the Third Law) is:

It is impossible for any process, no matter how idealized, to reduce the entropy of a system to its absolute-zero value in a finite number of operations.

Physically, the Nernst–Simon statement implies that it is impossible for any procedure to bring a system to the absolute zero of temperature in a finite number of steps.

History

The 3rd law was developed by the chemist Walther Nernst during the years 1906–12, and is therefore often referred to as Nernst's theorem or Nernst's postulate. The third law of thermodynamics states that the entropy of a system at absolute zero is a well-defined constant. This is because a system at zero temperature exists in its ground state, so that its entropy is determined only by the degeneracy of the ground state.

In 1912 Nernst stated the law thus: "It is impossible for any procedure to lead to the isotherm $T = 0$ in a finite number of steps."

An alternative version of the third law of thermodynamics as stated by Gilbert N. Lewis and Merle Randall in 1923:

> If the entropy of each element in some (perfect) crystalline state be taken as zero at the absolute zero of temperature, every substance has a finite positive entropy; but at the absolute zero of temperature the entropy may become zero, and does so become in the case of perfect crystalline substances.

This version states not only ΔS will reach zero at 0 K, but S itself will also reach zero as long as the crystal has a ground state with only one configuration. Some crystals form defects which causes a residual entropy. This residual entropy disappears when the kinetic barriers to transitioning to one ground state are overcome.

With the development of statistical mechanics, the third law of thermodynamics (like the other laws) changed from a *fundamental* law (justified by experiments) to a *derived* law (derived from even more basic laws). The basic law from which it is primarily derived is the statistical-mechanics definition of entropy for a large system:

$$S - S_0 = k_B \ln \Omega$$

where S is entropy, k_B is the Boltzmann constant, and Ω is the number of microstates consistent with the macroscopic configuration. The counting of states is from the reference state of absolute zero, which corresponds to the entropy of S_0.

Explanation

In simple terms, the third law states that the entropy of a perfect crystal of a pure substance approaches zero as the temperature approaches zero. The alignment of a perfect crystal leaves no ambiguity as to the location and orientation of each part of the crystal. As the energy of the crystal is reduced, the vibrations of the individual atoms are reduced to nothing, and the crystal becomes the same everywhere.

The third law provides an absolute reference point for the determination of entropy at any other temperature. The entropy of a system, determined relative to this zero point, is then the *absolute* entropy of that system. Mathematically, the absolute entropy of any system at zero temperature is the natural log of the number of ground states times Boltzmann's constant $k_B = 1.38 \times 10^{-23}$, JK^{-1}.

The entropy of a *perfect* crystal lattice as defined by Nernst's theorem is zero provided that its ground state is unique, because $\ln(1) = 0$. If the system is composed of one-billion atoms, all alike, and lie within the matrix of a perfect crystal, the number of permutations of one-billion identical things taken one-billion at a time is $\Omega = 1$. Hence:

$$S - S_0 = k_B \ln \Omega = k_B \ln 1 = 0$$

The difference is zero, hence the initial entropy S_0 can be any selected value so long as all other such calculations include that as the initial entropy. As a result the initial entropy value of zero is selected $S_0 = 0$ is used for convenience.

$$S - S_0 = S - 0 = 0$$

By way of example, suppose a system consists of 1 cm³ of matter with a mass of 1 g and 20 g/mol. The system consists of 3×10^{22} identical atoms at 0 K. If one atom should absorb a photon of wavelength of 1 cm that atom is then unique and the permutations of one unique atom among the 3×10^{22} is $N = 3 \times 10^{22}$. The entropy, energy, and temperature of the system rises and can be calculated. The entropy change is:

$$\ddot{A}S = S - S_0 = k_B \ln \grave{U}$$

From the second law of thermodynamics:

$$\Delta S = S - S_0 = \frac{\delta Q}{T}$$

Hence:

$$\Delta S = S - S_0 = k_B \ln(\Omega) = \frac{\delta Q}{T}$$

Calculating entropy change:

$$S - 0 = k_B \ln N = 1.38 \times 10^{-23} \times \ln 3 \times 10^{22} = 70 \times 10^{-23} \, \mathrm{JK^{-1}}$$

The energy change of the system as a result of absorbing the single photon whose energy is ϵ:

$$\delta Q = \epsilon = \frac{hc}{\lambda} = \frac{6.62 \times 10^{-34} \, \mathrm{J \cdot s} \times 3 \times 10^8 \, \mathrm{ms^{-1}}}{0.01 \mathrm{m}} = 2 \times 10^{-23} \, \mathrm{J}$$

The temperature of the system rises by:

$$T = \frac{\epsilon}{\Delta S} = \frac{2 \times 10^{-23} \, \mathrm{J}}{70 \times 10^{-23} \, \mathrm{JK^{-1}}} = \frac{1}{35} \mathrm{K}$$

This can be interpreted as the average temperature of the system over the range from $0 < S < 70 \times 10^{-23}$ J/K A single atom was assumed to absorb the photon but the temperature and entropy change characterizes the entire system.

An example of a system which does not have a unique ground state is one whose net spin is a half-integer, for which time-reversal symmetry gives two degenerate ground states. For such systems, the entropy at zero temperature is at least $k_B *\ln(2)$ (which is negligible on a macroscopic scale). Some crystalline systems exhibit geometrical frustration, where the structure of the crystal lattice prevents the emergence of a unique ground state. Ground-state helium (unless under pressure) remains liquid.

In addition, glasses and solid solutions retain large entropy at 0 K, because they are large collections of nearly degenerate states, in which they become trapped out of equilibrium. Another example of a solid with many nearly-degenerate ground states, trapped out of equilibrium, is ice Ih, which has "proton disorder".

For the entropy at absolute zero to be zero, the magnetic moments of a perfectly ordered crystal must themselves be perfectly ordered; from an entropic perspective, this can be considered to be part of the definition of a "perfect crystal". Only ferromagnetic, antiferromagnetic, and diamagnetic materials can satisfy this condition. However, ferromagnetic materials do not in fact have zero entropy at zero temperature, because the spins of the unpaired electrons are all aligned and this gives a ground-state spin degeneracy. Materials that remain paramagnetic at 0 K, by contrast, may have many nearly-degenerate ground states (for example, in a spin glass), or may retain dynamic disorder (a quantum spin liquid).

Mathematical Formulation

Consider a closed system in internal equilibrium. As the system is in equilibrium, there are no irreversible processes so the entropy production is zero. During slow heating, small temperature gradients are generated in the material, but the associated entropy production can be kept arbitrarily low if the heat is supplied slowly enough. The increase in entropy due to the added heat δQ is then given by the second part of the Second law of thermodynamics which states that the entropy change of a system ΔS is given by

$$\Delta S = \frac{\delta Q}{T}. \tag{1}$$

The temperature rise dT due to the heat δQ is determined by the heat capacity $C(T,X)$ according to

$$\delta Q = C(T,X)dT. \tag{2}$$

The parameter X is a symbolic notation for all parameters (such as pressure, magnetic field, liquid/solid fraction, etc.) which are kept constant during the heat supply. E.g. if the volume is constant we get the heat capacity at constant volume C_V. In the case of a phase transition from liquid to solid, or from gas to liquid the parameter X can be one of the two components. Combining relations (1) and (2) gives

$$\Delta S = \frac{C(T,X)dT}{T}. \tag{3}$$

Integration of Eq.(3) from a reference temperature T_0 to an arbitrary temperature T

gives the entropy at temperature T

$$S(T,X) = S(T_0,X) + \int_{T_0}^{T} \frac{C(T',X)}{T'} dT'. \tag{4}$$

We now come to the mathematical formulation of the third law. There are three steps:

1: in the limit $T_0 \to 0$ the integral in Eq.(4) is finite.So that we may take $T_0 = 0$ and write

$$S(T,X) = S(0,X) + \int_{0}^{T} \frac{C(T',X)}{T'} dT'. \tag{5}$$

2. the value of $S(0,X)$ is independent of X.In mathematical form

$$S(0,X) = S(0). \tag{6}$$

So Eq.(5) can be further simplified to

$$S(T,X) = S(0) + \int_{0}^{T} \frac{C(T',X)}{T'} dT'. \tag{7}$$

Equation (6) can also be formulated as

$$\lim_{T \to 0} \left(\frac{\partial S(T,X)}{\partial X} \right)_{T} = 0. \tag{8}$$

In words: at absolute zero all isothermal processes are isentropic. Eq.(8) is the mathematical formulation of the third law.

3: Classically, one is free to choose the zero of the entropy, and it is convenient to take

$$S(0) = 0 \tag{9}$$

so that Eq.(7) reduces to the final form

$$S(T,X) = \int_{0}^{T} \frac{C(T',X)}{T'} dT'. \tag{10}$$

However, reinterpreting Eq. (9) in view of the quantized nature of the lowest-lying energy states, the physical meaning of Eq.(9) goes deeper than just a convenient selection of the zero of the entropy. It is due to the perfect order at zero kelvin as explained above.

Consequences of the third Law

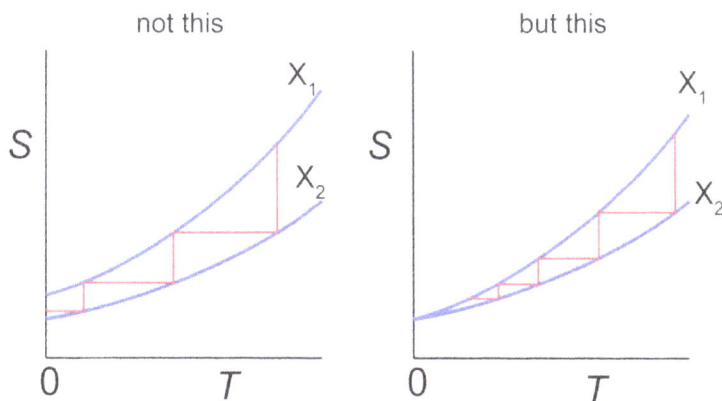

Fig.1 Left side: Absolute zero can be reached in a finite number of steps if $S(0,X_1) \neq S(0, X_2)$. Right: An infinite number of steps is needed since $S(0,X_1) = S(0,X_2)$.

Absolute Zero

The third law is equivalent to the statement that

> "It is impossible by any procedure, no matter how idealized, to reduce the temperature of any system to zero temperature in a finite number of finite operations".

The reason that $T=0$ cannot be reached according to the third law is explained as follows: Suppose that the temperature of a substance can be reduced in an isentropic process by changing the parameter X from X_2 to X_1. One can think of a multistage nuclear demagnetization setup where a magnetic field is switched on and off in a controlled way. If there were an entropy difference at absolute zero, $T=0$ could be reached in a finite number of steps. However, at $T=0$ there is no entropy difference so an infinite number of steps would be needed. The process is illustrated in Fig.1.

Specific Heat

A non-quantitative description of his third law that Nernst gave at the very beginning was simply that the specific heat can always be made zero by cooling the material down far enough. A modern, quantitative analysis follows.

Suppose that the heat capacity of a sample in the low temperature region has the form of a power law $C(T,X)=C_0 T^\alpha$ asymptotically as $T \to 0$, and we wish to find which values of α are compatible with the third law. We have

$$\int_{T_0}^{T} \frac{C(T',X)}{T'}dT' = \frac{C_0}{\alpha}(T^\alpha - T_0^\alpha).$$ (11)

By the discussion of third law (above), this integral must be bounded as $T_0 \to 0$, which is only possible if $\alpha > 0$. So the heat capacity must go to zero at absolute zero

$$\lim_{T \to 0} C(T,X) = 0.$$ (12)

if it has the form of a power law. The same argument shows that it cannot be bounded below by a positive constant, even if we drop the power-law assumption.

On the other hand, the molar specific heat at constant volume of a monatomic classical ideal gas, such as helium at room temperature, is given by $C_V=(3/2)R$ with R the molar ideal gas constant. But clearly a constant heat capacity does not satisfy Eq. (12). That is, a gas with a constant heat capacity all the way to absolute zero violates the third law of thermodynamics. We can verify this more fundamentally by substituting C_V in Eq. (4), which yields

$$S(T,V) = S(T_0,V) + \frac{3}{2}R\ln\frac{T}{T_0}.$$ (13)

In the limit $T_0 \to 0$ this expression diverges, again contradicting the third law of thermodynamics.

The conflict is resolved as follows: At a certain temperature the quantum nature of matter starts to dominate the behavior. Fermi particles follow Fermi–Dirac statistics and Bose particles follow Bose–Einstein statistics. In both cases the heat capacity at low temperatures is no longer temperature independent, even for ideal gases. For Fermi gases

$$C_V = \frac{\pi^2}{2}R\frac{T}{T_F}$$ (14)

with the Fermi temperature T_F given by

$$T_F = \frac{1}{8\pi^2}\frac{N_A^2 h^2}{MR}\left(\frac{3\pi^2 N_A}{V_m}\right)^{2/3}.$$ (15)

Here N_A is Avogadro's number, V_m the molar volume, and M the molar mass.

For Bose gases

$$C_V = 1.93..R\left(\frac{T}{T_B}\right)^{3/2} \tag{16}$$

with T^B given by

$$T_B = \frac{1}{11.9..}\frac{N_A^2 h^2}{MR}\left(\frac{N_A}{V_m}\right)^{2/3}. \tag{17}$$

The specific heats given by Eq.(14) and (16) both satisfy Eq.(12). Indeed, they are power laws with $\alpha=1$ and $\alpha=3/2$ respectively.

Vapor Pressure

The only liquids near absolute zero are ³He and ⁴He. Their heat of evaporation has a limiting value given by

$$L = L_0 + C_p T \tag{18}$$

with L_o and C_p constant. If we consider a container, partly filled with liquid and partly gas, the entropy of the liquid–gas mixture is

$$S(T,x) = S_l(T) + x\left(\frac{L_0}{T} + C_p\right) \tag{19}$$

where $S_l(T)$ is the entropy of the liquid and x is the gas fraction. Clearly the entropy change during the liquid–gas transition (x from 0 to 1) diverges in the limit of $T\to0$. This violates Eq.(8). Nature solves this paradox as follows: at temperatures below about 50 mK the vapor pressure is so low that the gas density is lower than the best vacuum in the universe. In other words: below 50 mK there is simply no gas above the liquid.

Latent Heat of Melting

The melting curves of ³He and ⁴He both extend down to absolute zero at finite pressure. At the melting pressure liquid and solid are in equilibrium. The third law demands that the entropies of the solid and liquid are equal at $T=0$. As a result the latent heat of melting is zero and the slope of the melting curve extrapolates to zero as a result of the Clausius–Clapeyron equation.

Thermal Expansion Coefficient

The thermal expansion coefficient is defined as

$$\alpha_V = \frac{1}{V_m}\left(\frac{\partial V_m}{\partial T}\right)_p. \tag{20}$$

With the Maxwell relation

$$\left(\frac{\partial V_m}{\partial T}\right)_p = -\left(\frac{\partial S_m}{\partial p}\right)_T \tag{21}$$

and Eq.(8) with $X=p$ it is shown that

$$\lim_{T\to 0}\alpha_V = 0. \tag{22}$$

So the thermal expansion coefficient of all materials must go to zero at zero kelvin.

References

- Truesdell, C., Muncaster, R.G. (1980). Fundamentals of Maxwell's Kinetic Theory of a Simple Monatomic Gas, Treated as a Branch of Rational Mechanics, Academic Press, New York, ISBN 0-12-701350-4, p. 15.

- Denbigh, K.G., Denbigh, J.S. (1985). Entropy in Relation to Incomplete Knowledge, Cambridge University Press, Cambridge UK, ISBN 0-521-25677-1, pp. 43–44.

- Grandy, W.T., Jr (2008). Entropy and the Time Evolution of Macroscopic Systems, Oxford University Press, Oxford, ISBN 978-0-19-954617-6, pp. 55–58.

- Adkins, C.J. (1968/1983). Equilibrium Thermodynamics, (1st edition 1968), third edition 1983, Cambridge University Press, Cambridge UK, ISBN 0-521-25445-0.

- Atkins, P.W., de Paula, J. (2006). Atkins' Physical Chemistry, eighth edition, W.H. Freeman, New York, ISBN 978-0-7167-8759-4.

- Attard, P. (2012). Non-equilibrium Thermodynamics and Statistical Mechanics: Foundations and Applications, Oxford University Press, Oxford UK, ISBN 978-0-19-966276-0.

- Callen, H.B. (1960/1985). Thermodynamics and an Introduction to Thermostatistics, (1st edition 1960) 2nd edition 1985, Wiley, New York, ISBN 0-471-86256-8.

- Čápek, V., Sheehan, D.P. (2005). Challenges to the Second Law of Thermodynamics: Theory and Experiment, Springer, Dordrecht, ISBN 1-4020-3015-0.

- Carnot, S. (1824/1986). Reflections on the motive power of fire, Manchester University Press, Manchester UK, ISBN 0-7190-1741-6.

- Denbigh, K. (1954/1981). The Principles of Chemical Equilibrium. With Applications in Chemistry and Chemical Engineering, fourth edition, Cambridge University Press, Cambridge UK, ISBN 0-521-23682-7.

- Griem, H.R. (2005). Principles of Plasma Spectroscopy (Cambridge Monographs on Plasma Physics), Cambridge University Press, New York ISBN 0-521-61941-6.

- Glansdorff, P., Prigogine, I. (1971). Thermodynamic Theory of Structure, Stability, and Fluctuations, Wiley-Interscience, London, 1971, ISBN 0-471-30280-5.

- Grandy, W.T., Jr (2008). Entropy and the Time Evolution of Macroscopic Systems. Oxford University Press. ISBN 978-0-19-954617-6.

- Kondepudi, D., Prigogine, I. (1998). Modern Thermodynamics: From Heat Engines to Dissipative Structures, John Wiley & Sons, Chichester, ISBN 0-471-97393-9.

- Münster, A. (1970), Classical Thermodynamics, translated by E.S. Halberstadt, Wiley–Interscience, London, ISBN 0-471-62430-6.

- Uffink, J. (2003). Irreversibility and the Second Law of Thermodynamics, Chapter 7 of Entropy, Greven, A., Keller, G., Warnecke (editors) (2003), Princeton University Press, Princeton NJ, ISBN 0-691-11338-6.

- Balescu, R. (1975). Equilibrium and Non-equilibrium Statistical Mechanics, Wiley-Interscience, New York, ISBN 0-471-04600-0, Section 3.2, pages 64-72.

- Glansdorff, P., Prigogine, I. (1971). Thermodynamic Theory of Structure, Stability, and Fluctuations, Wiley-Interscience, London, 1971, ISBN 0-471-30280-5.

- Jou, D., Casas-Vázquez, J., Lebon, G. (1993). Extended Irreversible Thermodynamics, Springer, Berlin, ISBN 3-540-55874-8, ISBN 0-387-55874-8.

- Balescu, R. (1975). Equilibrium and Non-equilibrium Statistical Mechanics, John Wiley & Sons, New York, ISBN 0-471-04600-0.

- Mihalas, D., Weibel-Mihalas, B. (1984). Foundations of Radiation Hydrodynamics, Oxford University Press, New York, ISBN 0-19-503437-6.

- Schloegl, F. (1989). Probability and Heat: Fundamentals of Thermostatistics, Freidr. Vieweg & Sohn, Brausnchweig, ISBN 3-528-06343-2.

- Keizer, J. (1987). Statistical Thermodynamics of Nonequilibrium Processes, Springer-Verlag, New York, ISBN 0-387-96501-7.

6

Thermal Insulation: A Comprehensive Study

Thermal insulation is achievable with the help of engineered methods or processes. It can also be achieved by using suitable objects with apt shapes and materials. The features of thermal insulation that have been covered in the following section are pipe insulation, building insulation and exhaust heat management. The chapter on thermal insulation offers an insightful focus, keeping in mind the complex topic.

Thermal Insulation

Mineral wool Insulation, 1600 dpi scan

Thermal insulation is the reduction of heat transfer (the transfer of thermal energy between objects of differing temperature) between objects in thermal contact or in range of radiative influence. Thermal insulation can be achieved with specially engineered methods or processes, as well as with suitable object shapes and materials.

Heat flow is an inevitable consequence of contact between objects of differing temperature. Thermal insulation provides a region of insulation in which thermal conduction is reduced or thermal radiation is reflected rather than absorbed by the lower-temperature body.

The insulating capability of a material is measured with thermal conductivity (k). Low thermal conductivity is equivalent to high insulating capability (R-value). In thermal engineering, other important properties of insulating materials are product density (ρ) and specific heat capacity (c).

Definition

Insulation

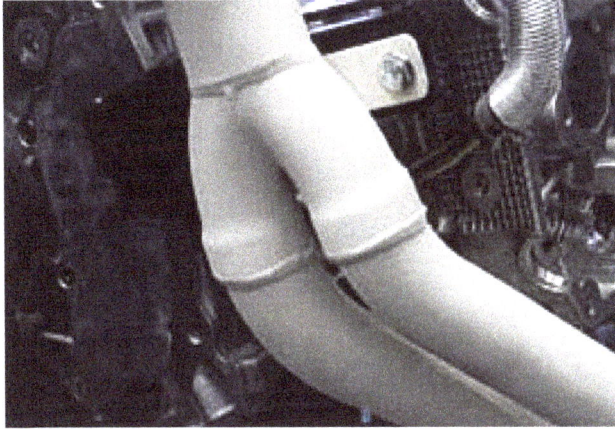

Car exhausts usually require some form of heat barrier, especially high performance exhausts where a ceramic coating is often applied

Solid materials chosen for insulation have a low thermal conductivity k, measured in watts-per-meter per kelvin (W·m⁻¹·K⁻¹). As the thickness of insulation is increased, the thermal resistance—or R-value—also increases.

For insulated cylinders, there is a complication – a *critical radius* beyond which extra insulation paradoxically increases heat transfer. The convective thermal resistance is inversely proportional to the surface area and therefore the radius of the cylinder, while the thermal resistance of a cylindrical shell (the insulation layer) depends on the ratio between outside and inside radius, not on the radius itself. If the outside radius of a cylinder is doubled by applying insulation, a fixed amount of conductive resistance (equal to $\ln(2)/(2\pi kL)$) is added. However, at the same time, the convective resistance has been halved. Because convective resistance tends to infinity when the radius approaches zero, at small enough radiuses the decrease in convective resistance will be larger than the added conductive resistance, resulting in lower total resistance. This implies that adding insulation actually increases the heat transfer, until a critical radius is reached, at which point the heat transfer is at maximum. Above this critical radius, added insulation decreases the heat transfer. For insulated cylinders, the critical radius is given by the equation

$$r_{critical} = \frac{k}{h}$$

This equation shows that the critical radius depends only on the heat transfer coefficient and the thermal conductivity of the insulation. If the radius of the uninsulated cylinder is larger than the critical radius for insulation, the addition of any amount of insulation will decrease heat transfer.

Applications

Clothing and Natural Animal Insulation in Birds and Mammals

Gases possess poor thermal conduction properties compared to liquids and solids, and thus makes a good insulation material if they can be trapped. In order to further augment the effectiveness of a gas (such as air) it may be disrupted into small cells which cannot effectively transfer heat by natural convection. Convection involves a larger bulk flow of gas driven by buoyancy and temperature differences, and it does not work well in small cells where there is little density difference to drive it.

In order to accomplish gas cell formation in man-made thermal insulation, glass and polymer materials can be used to trap air in a foam-like structure. This principle is used industrially in building and piping insulation such as (glass wool), cellulose, rock wool, polystyrene foam (styrofoam), urethane foam, vermiculite, perlite, and cork. Trapping air is also the principle in all highly insulating clothing materials such as wool, down feathers and fleece.

The air-trapping property is also the insulation principle employed by homeothermic animals to stay warm, for example down feathers, and insulating hair such as natural sheep's wool. In both cases the primary insulating material is air, and the polymer used for trapping the air is natural keratin protein.

Buildings

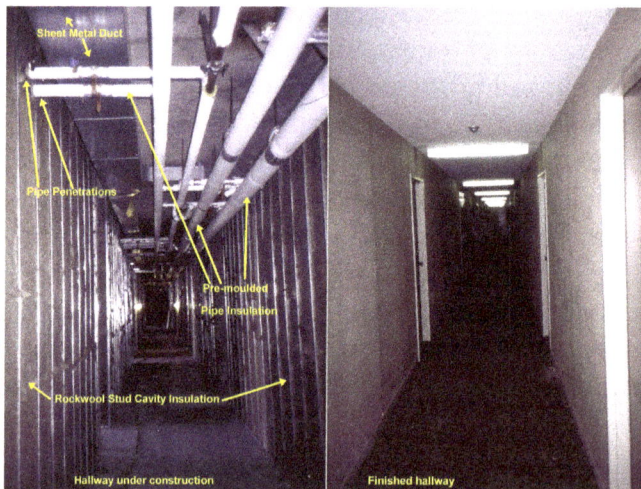

Common insulation applications in apartment building in Ontario, Canada.

Maintaining acceptable temperatures in buildings (by heating and cooling) uses a large proportion of global energy consumption. Building insulations also commonly use the principle of small trapped air-cells as explained above, e.g. fiberglass (specifically glass wool), cellulose, rock wool, polystyrene foam, urethane foam, vermiculite, perlite, cork, etc. For a period of time, Asbestos was also used, however, it caused health problems.

When well insulated, a building:

- is energy-efficient, thus saving the owner money.

- provides more uniform temperatures throughout the space. There is less temperature gradient both vertically (between ankle height and head height) and horizontally from exterior walls, ceilings and windows to the interior walls, thus producing a more comfortable occupant environment when outside temperatures are extremely cold or hot.

- has minimal recurring expense. Unlike heating and cooling equipment, insulation is permanent and does not require maintenance, upkeep, or adjustment.

- lowers the carbon footprint of a building.

Many forms of thermal insulation also reduce noise and vibration, both coming from the outside and from other rooms inside a building, thus producing a more comfortable environment.

Window insulation film can be applied in weatherization applications to reduce incoming thermal radiation in summer and loss in winter.

In industry, energy has to be expended to raise, lower, or maintain the temperature of objects or process fluids. If these are not insulated, this increases the energy requirements of a process, and therefore the cost and environmental impact.

Mechanical Systems

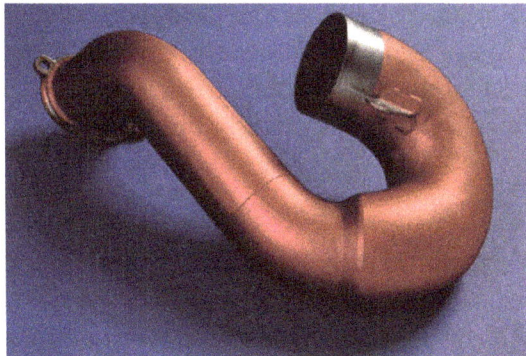

Thermal insulation applied to exhaust component by means of plasma spraying

Space heating and cooling systems distribute heat throughout buildings by means of pipe or ductwork. Insulating these pipes using pipe insulation reduces energy into unoccupied rooms and prevents condensation from occurring on cold and chilled pipework.

Pipe insulation is also used on water supply pipework to help delay pipe freezing for an acceptable length of time.

Spacecraft

Thermal insulation on the Huygens probe

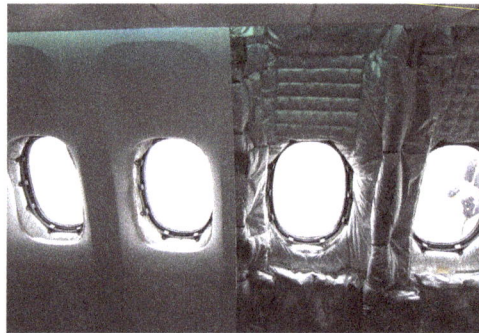

Cabin insulation of a Boeing 747-8 airliner

Launch and re-entry place severe mechanical stresses on spacecraft, so the strength of an insulator is critically important (as seen by the failure of insulating foam on the Space Shuttle Columbia). Re-entry through the atmosphere generates very high temperatures due to compression of the air at high speeds. Insulators must meet demanding physical properties beyond their thermal transfer retardant properties. E.g. reinforced carbon-carbon composite nose cone and silica fiber tiles of the Space Shuttle.

Automotive

Internal combustion engines produce a lot of heat during their combustion cycle. This can have a negative effect when it reaches various heat-sensitive components such as sensors, batteries and starter motors. As a result, thermal insulation is necessary to prevent the heat from the exhaust reaching these components.

High performance cars often use thermal insulation as a means to increase engine performance.

Factors Influencing Performance

Insulation performance is influenced by many factors the most prominent of which include:

- Thermal conductivity ("k" or "λ" value)

- Surface emissivity ("ε" value)

- Insulation thickness

- Density

- Specific heat capacity

- Thermal bridging

It is important to note that the factors influencing performance may vary over time as material ages or environmental conditions change.

Calculating Requirements

Industry standards are often rules of thumb, developed over many years, that offset many conflicting goals: what people will pay for, manufacturing cost, local climate, traditional building practices, and varying standards of comfort. Both heat transfer and layer analysis may be performed in large industrial applications, but in household situations (appliances and building insulation), air tightness is the key in reducing heat transfer due to air leakage (forced or natural convection). Once air tightness is achieved, it has often been sufficient to choose the thickness of the insulating layer based on rules of thumb. Diminishing returns are achieved with each successive doubling of the insulating layer. It can be shown that for some systems, there is a minimum insulation thickness required for an improvement to be realized.

Pipe Insulation

Pipe Insulation is thermal or acoustic insulation used on pipework.

Applications

Condensation Control

Where pipes operate at below-ambient temperatures, the potential exists for water vapour to condense on the pipe surface. Moisture is known to contribute towards many different types of corrosion, so preventing the formation of condensation on pipework is usually considered important.

Pipe insulation can prevent condensation forming, as the surface temperature of the insulation will vary from the surface temperature of the pipe. Condensation will not occur, provided that (a) the insulation surface is above the dewpoint temperature of the air; and (b) the insulation incorporates some form of water-vapour barrier or retarder that prevents water vapour from passing through the insulation to form on the pipe surface. to preventing the corrosion on pipe line, insulation are required with proper thickness and density.

Pipe Freezing

Since some water pipes are located either outide or in unheated areas where the ambient temperature may occasionally drop below the freezing point of water, any water in the pipework may potentially freeze. When water freezes, it expands due to negative thermal expansion, and this expansion can cause failure of a pipe system in any one of a number of ways.

Pipe insulation cannot prevent the freezing of standing water in pipework, but it can increase the time required for freezing to occur—thereby reducing the risk of the water in the pipes freezing. For this reason, it is recommended to insulate pipework at risk of freezing, and local water-supply regulations may require pipe insulation be applied to pipework to reduce the risk of pipe freezing.

For a given length, a smaller-bore pipe holds a smaller volume of water than a larger-bore pipe, and therefore water in a smaller-bore pipe will freeze more easily (and more quickly) than water in a larger-bore pipe (presuming equivalent environments). Since smaller-bore pipes present a greater risk of freezing, insulation is typically used in combination with alternative methods of freeze prevention (e.g., modulating trace heating cable, or ensuring a consistent flow of water through the pipe).

Energy Saving

Since pipework can operate at temperatures far removed from the ambient temperature, and the rate of heat flow from a pipe is related to the temperature differential between the pipe and the surrounding ambient air, heat flow from pipework can be considerable. In many situations, this heat flow is undesirable. The application of thermal pipe insulation introduces thermal resistance and reduces the heat flow.

Thicknesses of thermal pipe insulation used for saving energy vary, but as a general rule, pipes operating at more-extreme temperatures exhibit a greater heat flow and larger thicknesses are applied due to the greater potential savings.

the different size of pipe related with the energy saving. Depend on the flow rate, energy cost is determined by particular pipe size. Using PIPEOPT algorithm, if pipe diameter is lager, the energy is less than smaller pipe diameter because of the lower pressure.

The location of pipework also influences the selection of insulation thickness. For instance, in some circumstances, heating pipework within a well-insulated building might not require insulation, as the heat that's "lost" (i.e., the heat that flows from the pipe to the surrounding air) may be considered "useful" for heating the building, as such "lost" heat would be effectively trapped by the structural insulation anyway. Conversely, such pipework may be insulated to prevent overheating or unnecessary cooling in the rooms through which it passes.

Protection Against Extreme Temperatures

Where pipework is operating at extremely high or low temperatures, the potential exists for injury to occur should any person come into physical contact with the pipe surface. The threshold for human pain varies, but several international standards set recommended touch temperature limits.

Since the surface temperature of insulation varies from the temperature of the pipe surface, typically such that the insulation surface has a "less extreme" temperature, pipe insulation can be used to bring surface touch temperatures into a safe range.

Control of Noise

Pipework can operate as a conduit for noise to travel from one part of a building to another (a typical example of this can be seen with waste-water pipework routed within a building). Acoustic insulation can prevent this noise transfer by acting to damp the pipe wall and performing an acoustic decoupling function wherever the pipe passes through a fixed wall or floor and wherever the pipe is mechanically fixed.

Pipework can also radiate mechanical noise. In such circumstances, the breakout of noise from the pipe wall can be achieved by acoustic insulation incorporating a high-density sound barrier.

Factors Influencing Performance

The relative performance of different pipe insulation on any given application can be influenced by many factors. The principal factors are:

- Thermal conductivity ("k" or "λ" value)

- Surface emissivity ("ε" value)

- Water-vapour resistance ("μ" value)

- Insulation thickness

- Density

Other factors, such as the level of moisture content and the opening of joints, can influence the overall performance of pipe insulation. Many of these factors are listed in the international standard EN ISO 23993.

Materials

Pipe insulation materials come in a large variety of forms, but most materials fall into one of the following categories.

Mineral Wool

Mineral wools, including rock and slag wools, are inorganic strands of mineral fibre bonded together using organic binders. Mineral wools are capable of operating at high temperatures and exhibit good fire performance ratings when tested.

Mineral wools are used on all types of pipework, particularly industrial pipework operating at higher temperatures.

Glass Wool

Glass wool is a high-temperature fibrous insulation material, similar to mineral wool, where inorganic strands of glass fibre are bound together using a binder.

As with other forms of mineral wool, glass-wool insulation can be used for thermal and acoustic applications.

Flexible Elastomeric Foams

These are flexible, closed-cell, rubber foams based on NBR or EPDM rubber. Flexible elastomeric foams exhibit such a high resistance to the passage of water vapour that they do not generally require additional water-vapour barriers. Such high vapour resistance, combined with the high surface emissivity of rubber, allows flexible elastomeric foams to prevent surface condensation formation with comparatively small thicknesses.

As a result, flexible elastomeric foams are widely used on refrigeration and air-conditioning pipework. Flexible elastomeric foams are also used on heating and hot-water systems.

Rigid Foam

Pipe insulation made from rigid Phenolic, PIR, or PUR foam insulation is common in some countries. Rigid-foam insulation has minimal acoustic performance but can exhibit low thermal-conductivity values of 0.021 W/(m·K) or lower, allowing energy-saving legislation to be met whilst using reduced insulation thicknesses.

Polyethylene

Polyethylene is a flexible plastic foamed insulation that is widely used to prevent freezing of domestic water supply pipes and to reduce heat loss from domestic heating pipes.

The fire performance of Polyethylene is typically 25/50 E84 compliant up to 1" thickness.

Cellular Glass

100% Glass manufactured primarily from sand, limestone & soda ash..

Aerogel

Silica Aerogel insulation has the lowest thermal conductivity of any commercially produced insulation. Although no manufacturer currently manufactures Aerogel pipe sections, it is possible to wrap Aerogel blanket around pipework, allowing it to function as pipe insulation.

The usage of Aerogel for pipe insulation is currently limited.

Heat Flow calculations and R-value

Heat flow passing through pipe insulation can be calculated by following the equations set out in either the ASTM C 680 or EN ISO 12241 standards. Heat flow is given by the following equation:

$$q = \frac{\Theta_i - \Theta_a}{R_T}$$

Where:

- Θ_i is the internal pipe temperature,

- Θ_a is the external ambient temperature, and

- R_T is the sum total thermal resistance of all insulation layers and the internal- and external-surface heat-transfer resistances.

In order to calculate heat flow, it is first necessary to calculate the thermal resistance ("R-value") for each layer of insulation.

For pipe insulation, the R-value varies not only with the insulation thickness and thermal conductivity ("k-value") but also with the pipe outer diameter and the average material temperature. For this reason, it is more common to use the thermal conductivity value when comparing the effectiveness of pipe insulation, and R-values of pipe insulation are not covered by the US FTC R-value rule.

The thermal resistance of each insulation layer is calculated using the following equation:

$$R = \frac{D_x \ln(D_e / D_i)}{\lambda}$$

Where:

- D_e represents the insulation outer diameter,

- D_i represents the insulation inner diameter,

- λ represents the thermal conductivity ("k-value") at the average insulation temperature (for accurate results iterative calculations are necessary), and

- D_x is either D_e if heat loss calculation will use D_e for area calculation or D_i if it will use D_i.

Calculating the heat transfer resistance of the inner- and outer-insulation surfaces is more complex and requires the calculation of the internal- and external-surface coefficients of heat transfer. Equations for calculating this are based on empirical results and vary from standard to standard (both ASTM C 680 and EN ISO 12241 contain equations for estimating surface coefficients of heat transfer).

A number of organisations such as the North American Insulation Manufacturers Association and Firo Insulation offer free programs that allow the calculation of heat flow through pipe insulation.

Building Insulation

Mineral wool insulation

Building insulation refers broadly to any object in a building used as insulation for any purpose. While the majority of insulation in buildings is for thermal purposes, the term also applies to acoustic insulation, fire insulation, and impact insulation (e.g. for vibrations caused by industrial applications). Often an insulation material will be chosen for its ability to perform several of these functions at once.

Thermal Insulation

Thermal insulation in buildings is an important factor to achieving thermal comfort for its occupants. Insulation reduces unwanted heat loss or gain and can decrease the energy demands of heating and cooling systems. It does not necessarily deal with issues of adequate ventilation and may or may not affect the level of sound insulation. In a narrow sense insulation can just refer to the insulation materials employed to slow heat loss, such as: cellulose, glass wool, rock wool, polystyrene, urethane foam, vermiculite, perlite, wood fibre, plant fibre (cannabis, flax, cotton, cork, etc.), recycled cotton denim, plant straw, animal fibre (sheep's wool), cement, and earth or soil, Reflective Insulation (also known as Radiant Barrier) but it can also involve a range of designs and techniques to address the main modes of heat transfer - conduction, radiation and convection materials. Many of the materials in this list deal with heat conduction and convection by the simple expedient of trapping large amounts of air (or other gas) in a way that results in a material that employs the low thermal conductivity of small pockets of gas, rather than the much higher conductivity of typical solids. (A similar gas-trapping principle is used in animal hair, down feathers, and in air-containing insulating fabrics).

The effectiveness of Reflective Insulation (Radiant Barrier) is commonly evaluated by the Reflectivity (Emittance) of the surface with airspace facing to the heat source.

The effectiveness of bulk insulation is commonly evaluated by its R-value, of which there are two - metric (SI) and US customary, the former being 0.176 times the latter. For attics, it is recommended that it should be at least R-38 (US customary, R-6.7 metric). However, an R-value does not take into account the quality of construction or local environmental factors for each building. Construction quality issues include inadequate vapor barriers, and problems with draft-proofing. In addition, the properties and density of the insulation material itself is critical.

Planning

How much insulation a house should have depends on building design, climate, energy costs, budget, and personal preference. Regional climates make for different requirements. Building codes specify only the bare minimum; insulating beyond what the code requires is often recommended.

The insulation strategy of a building needs to be based on a careful consideration of the mode of energy transfer and the direction and intensity in which it moves. This may alter throughout the day and from season to season. It is important to choose an appropriate design, the correct combination of materials and building techniques to suit the particular situation.

In the USA

To determine whether you should add insulation, you first need to find out how much insulation you already have in your home and where. A qualified home energy auditor will include an insulation check as a routine part of a whole-house energy audit. However, you can sometimes perform a self-assessment in certain areas of the home, such as attics. Here, a visual inspection, along with use of a ruler, can give you a sense of whether you may benefit from additional insulation.

An initial estimate of insulation needs in the United States can be determined by the US Department of Energy's ZIP code insulation calculator.

Russia

In Russia the luxury of cheap gas has led to poorly insulated, overheated and inefficient consumers of energy. The Russian Center for Energy Efficiency found that Russian buildings are either over- or under-heated, and often consume up to fifty percent more heat and hot water than needed. Fifty-three percent of all carbon dioxide (CO_2) emissions in Russia are produced through heating and generating electricity for buildings. However, greenhouse gas emissions from the former Soviet Bloc are still below their 1990 levels.

Cold Climates

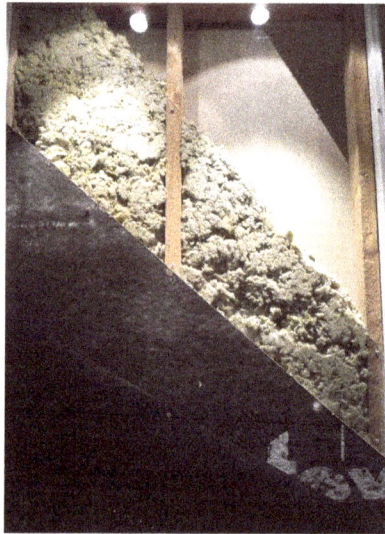

Cross-section of home insulation.

In cold conditions, the main aim is to reduce heat flow out of the building. The components of the building envelope - windows, doors, roofs, walls, and air infiltration barriers - are all important sources of heat loss; in an otherwise well insulated home, windows will then become an important source of heat transfer. The resistance to conducted heat loss for standard glazing corresponds to an R-value of about 0.17 m²K°/W (compared to 2-4 m²K°/W for glass wool batts). Losses can be reduced by good weatherisation, bulk insulation, and minimising the amount of non-insulative (particularly non-solar facing) glazing. Indoor thermal radiation can also be a disadvantage with spectrally selective (low-e, low-emissivity) glazing. Some insulated glazing systems can double to triple R values.

Hot Climates

In hot conditions, the greatest source of heat energy is solar radiation. This can enter buildings directly through windows or it can heat the building shell to a higher temperature than the ambient, increasing the heat transfer through the building envelope. The Solar Heat Gain Co-efficient (SHGC) (a measure of solar heat transmittance) of standard single glazing can be around 78-85%. Solar gain can be reduced by adequate shading from the sun, light coloured roofing, spectrally selective (heat-reflective) paints and coatings and various types of insulation for the rest of the envelope. Specially coated glazing can reduce SHGC to around 10%. Radiant barriers are highly effective for attic spaces in hot climates. In this application, they are much more effective in hot climates than cold climates. For downward heat flow, convection is weak and radiation dominates heat transfer across an air space. Radiant barriers must face an adequate air-gap to be effective.

If refrigerative air-conditioning is employed in a hot, humid climate, then it is particularly important to seal the building envelope. Dehumidification of humid air infiltration can waste significant energy. On the other hand, some building designs are based on effective cross-ventilation instead of refrigerative air-conditioning to provide convective cooling from prevailing breezes.

Orientation - passive Solar Design

Optimal placement of building elements (e.g. windows, doors, heaters) can play a significant role in insulation by considering the impact of solar radiation on the building and the prevailing breezes. Reflective laminates can help reduce passive solar heat in pole barns, garages, and metal buildings.

Building Envelope

The thermal envelope defines the conditioned or living space in a house. The attic or basement may or may not be included in this area. Reducing airflow from inside to outside can help to reduce convective heat transfer significantly.

Ensuring low convective heat transfer also requires attention to building construction (weatherization) and the correct installation of insulative materials.

The less natural airflow into a building, the more mechanical ventilation will be required to support human comfort. High humidity can be a significant issue associated with lack of airflow, causing condensation, rotting construction materials, and encouraging microbial growth such as mould and bacteria. Moisture can also drastically reduce the effectiveness of insulation by creating a thermal bridge. Air exchange systems can be actively or passively incorporated to address these problems.

Thermal Bridge

Thermal bridges are points in the building envelope that allow heat conduction to occur. Since heat flows through the path of least resistance, thermal bridges can contribute to poor energy performance. A thermal bridge is created when materials create a continuous path across a temperature difference, in which the heat flow is not interrupted by thermal insulation. Common building materials that are poor insulators include glass and metal.

A building design may have limited capacity for insulation in some areas of the structure. A common construction design is based on stud walls, in which thermal bridges are common in wood or steel studs and joists, which are typically fastened with metal. Notable areas that most commonly lack sufficient insulation are the corners of buildings, and areas where insulation has been removed or displaced to make room for system infrastructure, such as electrical boxes (outlets and light switches), plumbing, fire alarm equipment, etc.

Thermal bridges can also be created by uncoordinated construction, for example by closing off parts of external walls before they are fully insulated. The existence of inaccessible voids within the wall cavity which are devoid of insulation can be a source of thermal bridging.

Some forms of insulation transfer heat more readily when wet, and can therefore also form a thermal bridge in this state.

The heat conduction can be minimized by any of the following: reducing the cross sectional area of the bridges, increasing the bridge length, or decreasing the number of thermal bridges.

One method of reducing thermal bridge effects is the installation of an insulation board (e.g. foam board EPS XPS, wood fibre board, etc.) over the exterior outside wall. Another method is using insulated lumber framing for a thermal break inside the wall.

Installation

Insulating buildings during construction is much easier than retrofitting, as generally the insulation is hidden, and parts of the building need to be deconstructed to reach them.

Materials

There are essentially two types of building insulation - bulk insulation and reflective insulation. Most buildings use a combination of both types to make up a total building insulation system. The type of insulation used is matched to create maximum resistance to each of the three forms of building heat transfer - conduction, convection, and radiation.

Conductive and Convective Insulators

Bulk insulators block conductive heat transfer and convective flow either into or out of a building. The denser a material is, the better it will conduct heat. Because air has such low density, air is a very poor conductor and therefore makes a good insulator. Insulation to resist conductive heat transfer uses air spaces between fibers, inside foam or plastic bubbles and in building cavities like the attic. This is beneficial in an actively cooled or heated building, but can be a liability in a passively cooled building; adequate provisions for cooling by ventilation or radiation are needed.

Radiant Heat Barriers

Radiant barriers work in conjunction with an air space to reduce radiant heat transfer across the air space. Radiant or reflective insulation reflects heat instead of either absorbing it or letting it pass through. Radiant barriers are often seen used in reducing downward heat flow, because upward heat flow tends to be dominated by convection. This means that for attics, ceilings, and roofs, they are most effective in hot climates. They also have a role in reducing heat losses in cool climates. However, much greater insulation can be achieved through the addition of bulk insulators.

Some radiant barriers are spectrally selective and will preferentially reduce the flow of infra-red radiation in comparison to other wavelengths. For instance low-emissivity (low-e) windows will transmit light and short-wave infra-red energy into a building but reflect back the long-wave infra-red radiation generated by interior furnishings. Similarly, special heat-reflective paints are able to reflect more heat than visible light, or vice versa.

Thermal emissivity values probably best reflect the effectiveness of radiant barriers. Some manufacturers quote an 'equivalent' R-value for these products but these figures can be difficult to interpret, or even misleading, since R-value testing measures total heat loss in a laboratory setting and does not control the type of heat loss responsible for the net result (radiation, conduction, convection).

A film of dirt or moisture can alter the emissivity and hence the performance of radiant barriers.

Eco-friendly Insulation

Eco-friendly insulation is a term used for insulating products with limited environmental impact. The commonly accepted approach to determine whether or not an insulation products, but in fact any product or service is eco-friendly is by doing a life-cycle assessment (LCA). A number of studies compared the environmental impact of insulation materials in their application. The comparison shows that most important is the insulation value of the product meeting the technical requirements for the application. Only in a second order step a differentiation between materials becomes relevant. The report commissioned by the Belgian government to VITO is a good example of such a study. A valuable way to graphically represent such results is by a spider diagram.

Exhaust Heat Management

Exhaust Heat Management is the means of lessening the negative effects of internal combustion engine exhausts by preventing heat from escaping the exhaust system on automobiles.

Overview

The fact that exhausts often pass near important components means that ways of protecting these components from heat soak are especially important. As a result, heat management is used as a way of reflecting, dissipating or simply absorbing the heat.

Heat shield is also useful to reduce under-bonnet temperatures which therefore reduces intake temperature, subsequently increasing power.

Types

There are many different types of heat management, some more effective than others. They range from basic solid heat shield to plasma sprayed ceramics.

Heat Shield

Heat shield is one of the most widely used heat management options available due to its relative low price and ease to fit. In the past it has usually been custom made from rigid steel; however, flexible aluminium is now the standard. The key difference between a heat shield and insulating the pipe, through either wrapping or thermal coating, is the air gap that exists between the exhaust and the shield.

More recently technology has become available to apply ceramic thermal barrier coatings onto flexible aluminium in order to increase the thermal insulatory properties. This same procedure is also used on composite materials, which is often used on high-performance race cars, such as in Formula 1.

Ceramic Paint

This fairly low-cost solution can offer minor reductions in heat loss. It is often applied by brushing or spraying from an aerosol onto the exhaust system, followed by curing in the oven to allow the paint to adhere. This is sometimes applied to the internal surfaces of the exhaust. However, this often flakes prematurely due to the inability to clean the surface prior to application. It is usually necessary to reapply every 1–2 years due to it flaking and peeling.

Thermal Barrier Coating

These ceramic coatings are highly advanced coatings applied via plasma spray, and as a result, the coating is effectively welded to the surface of the exhaust system. They can offer small performance benefits due to the low thermal conductivity of the ceramic compound. This reduces engine bay temperature and increases exhaust velocity. These coatings protect steel exhaust systems further by protecting from rust.

Heat Wrap

This is the cheapest solution and is quite easy to apply. Exhaust heat wrapping has been used for many years to improve performance and avoid burns from motorcycle exhausts. Heat wrap consists of a high-temperature synthetic fabric which is wrapped around the manifold. Often sold as a cheap and easy way to boost horsepower, exhaust wrap does not increase engine output much. It can decrease engine bay temperatures and increase exhaust velocity. Lower intake air temperature and therefore increased air density, which may result in an increase in efficiency and power. Hotter exhaust gases travel faster which may reduce turbo lag and improve exhaust gas scavenging.,

Heat Wrap or Thermal Barrier Coating?

There has been much debate if heat wrap is better than plasma-sprayed coatings. The far cheaper option in the short term is heat wrap being quick and easy to apply. Heat wraps are more fragile and can absorb oil and water. Ceramic plasma-sprayed coatings also have similar performance benefits and whilst being quite expensive and requiring professional application, they last a long time and protect non stainless steel manifolds from rust.

References

- Hosier, Ralph (2013), BMW E30 3 Series: How to Modify for High-Performance and Competition, Veloce Publishing, p. 61, ISBN 1845844386

- "Archived copy" (PDF). Archived from the original (PDF) on 2011-10-05. Retrieved 2011-01-21. Pavatex Diffutherm ETICS

7

Theory of Thermal Equilibrium

The themes discussed in this section are thermodynamic system, thermodynamic process and thermodynamic equilibrium. This section helps the readers in developing an in-depth understanding of the theory of thermal equilibrium. The chapter strategically encompasses and incorporates the key concepts of thermal equilibrium, providing a complete understanding.

Thermodynamic System

A thermodynamic system is the material and radiative content of a macroscopic volume in space, that can be adequately described by thermodynamic state variables such as temperature, entropy, internal energy and pressure. Usually, by default, a thermodynamic system is taken to be in its own internal state of thermodynamic equilibrium, as opposed to a non-equilibrium state. The thermodynamic system is always enclosed by walls that separate it from its surroundings; these constrain the system. A thermodynamic system is subject to external interventions called thermodynamic operations; these alter the system's walls or its surroundings; as a result, the system undergoes thermodynamic processes according to the principles of thermodynamics. (This account mainly refers to the simplest kind of thermodynamic system; compositions of simple systems may also be considered.)

The thermodynamic state of a thermodynamic system is its internal state as specified by its state variables. In addition to the state variables, a thermodynamic account also requires a special kind of quantity called a state function, which is a function of the defining state variables. For example, if the state variables are internal energy, volume and mole amounts, that special function is the entropy. These quantities are inter-related by one or more functional relationships called equations of state, and by the system's characteristic equation. Thermodynamics imposes restrictions on the possible equations of state and on the characteristic equation. The restrictions are imposed by the laws of thermodynamics.

According to the permeabilities of the walls of a system, transfers of energy and matter occur between it and its surroundings, which are assumed to be unchanging over time, until a state of thermodynamic equilibrium is attained. The only states considered in equilibrium thermodynamics are equilibrium states. Classical thermodynamics includes equilibrium thermodynamics. It also considers: (a) systems considered in terms of cyclic sequences of processes rather than of states of the system; such were historically import-ant in the conceptual development of the subject; and (b) systems considered in terms of processes described by steady flows; such are important in engineering.

In 1824 Sadi Carnot described a thermodynamic system as the working substance (such as the volume of steam) of any heat engine under study. The very existence of such ther-modynamic systems may be considered a fundamental postulate of equilibrium ther-modynamics, though it is only rarely cited as a numbered law. According to Bailyn, the commonly rehearsed statement of the zeroth law of thermodynamics is a consequence of this fundamental postulate.

In equilibrium thermodynamics the state variables do not include fluxes because in a state of thermodynamic equilibrium all fluxes have zero values by postulation. Equi-librium thermodynamic processes may of course involve fluxes but these must have ceased by the time a thermodynamic process or operation is complete bringing a sys-tem to its eventual thermodynamic state. Non-equilibrium thermodynamics allows its state variables to include non-zero fluxes, that describe transfers of matter or energy or entropy between a system and its surroundings.

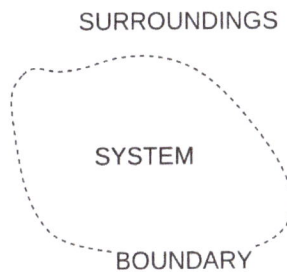

SURROUNDINGS

SYSTEM

BOUNDARY

Overview

Thermodynamic equilibrium is characterized by absence of flow of matter or energy. Equilibrium thermodynamics, as a subject in physics, considers macroscopic bodies of matter and energy in states of internal thermodynamic equilibrium. It uses the con-cept of thermodynamic processes, by which bodies pass from one equilibrium state to another by transfer of matter and energy between them. The term 'thermodynamic system' is used to refer to bodies of matter and energy in the special context of ther-modynamics. The possible equilibria between bodies are determined by the physical properties of the walls that separate the bodies. Equilibrium thermodynamics in gener-al does not measure time. Equilibrium thermodynamics is a relatively simple and well settled subject. One reason for this is the existence of a well defined physical quantity called 'the entropy of a body'.

Non-equilibrium thermodynamics, as a subject in physics, considers bodies of matter and energy that are not in states of internal thermodynamic equilibrium, but are usu-ally participating in processes of transfer that are slow enough to allow description in terms of quantities that are closely related to thermodynamic state variables. It is characterized by presence of flows of matter and energy. For this topic, very often the bodies considered have smooth spatial inhomogeneities, so that spatial gradients, for example a temperature gradient, are well enough defined. Thus the description of non-equilibrium thermodynamic systems is a field theory, more complicated than the theory of equilibrium thermodynamics. Non-equilibrium thermodynamics is a growing subject, not an established edifice. In general, it is not possible to find an exactly defined entropy for non-equilibrium problems. For many non-equilibrium thermodynamical problems, an approximately defined quantity called 'time rate of entropy production' is very useful. Non-equilibrium thermodynamics is mostly beyond the scope of the present article.

Another kind of thermodynamic system is considered in engineering. It takes part in a flow process. The account is in terms that approximate, well enough in practice in many cases, equilibrium thermodynamical concepts. This is mostly beyond the scope of the present article, and is set out in other articles, for example the article Flow process.

History

The first to create the concept of a thermodynamic system was the French physicist Sadi Carnot whose 1824 *Reflections on the Motive Power of Fire* studied what he called the *working substance*, e.g., typically a body of water vapor, in steam engines, in regards to the system's ability to do work when heat is applied to it. The working substance could be put in contact with either a heat reservoir (a boiler), a cold reservoir (a stream of cold water), or a piston (to which the working body could do work by pushing on it). In 1850, the German physicist Rudolf Clausius generalized this picture to include the concept of the surroundings, and began referring to the system as a "working body." In his 1850 manuscript *On the Motive Power of Fire*, Clausius wrote:

> 66 "With every change of volume (to the working body) a certain amount work must be done by the gas or upon it, since by its expansion it overcomes an external pressure, and since its compression can be brought about only by an exertion of external pressure. To this excess of work done by the gas or upon it there must correspond, by our principle, a proportional excess of heat consumed or produced, and the gas cannot give up to the "surrounding medium" the same amount of heat as it receives." 99

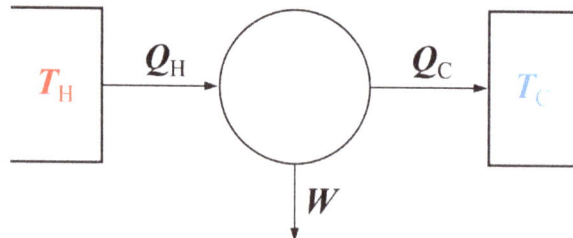

Carnot engine diagram (modern) - where heat flows from a high temperature T_H furnace through the fluid of the "working body" (working substance) and into the cold sink T_C, thus forcing the working substance to do mechanical work W on the surroundings, via cycles of contractions and expansions.

The article Carnot heat engine shows the original piston-and-cylinder diagram used by Carnot in discussing his ideal engine; below, we see the Carnot engine as is typically modeled in current use:

In the diagram shown, the "working body" (system), a term introduced by Clausius in 1850, can be any fluid or vapor body through which heat Q can be introduced or transmitted through to produce work. In 1824, Sadi Carnot, in his famous paper *Reflections on the Motive Power of Fire*, had postulated that the fluid body could be any substance capable of expansion, such as vapor of water, vapor of alcohol, vapor of mercury, a permanent gas, or air, etc. Though, in these early years, engines came in a number of configurations, typically Q_H was supplied by a boiler, wherein water boiled over a furnace; Q_C was typically a stream of cold flowing water in the form of a condenser located on a separate part of the engine. The output work W was the movement of the piston as it turned a crank-arm, which typically turned a pulley to lift water out of flooded salt mines. Carnot defined work as "weight lifted through a height."

Systems in Equilibrium

At thermodynamic equilibrium, a system's properties are, by definition, unchanging in time. Systems in equilibrium are much simpler and easier to understand than systems not in equilibrium. In some cases, when analyzing a thermodynamic process, one can assume that each intermediate state in the process is at equilibrium. This considerably simplifies the analysis.

In isolated systems it is consistently observed that as time goes on internal rearrangements diminish and stable conditions are approached. Pressures and temperatures tend to equalize, and matter arranges itself into one or a few relatively homogeneous phases. A system in which all processes of change have gone practically to completion is considered in a state of thermodynamic equilibrium. The thermodynamic properties of a system in equilibrium are unchanging in time. Equilibrium system states are much easier to describe in a deterministic manner than non-equilibrium states.

For a process to be reversible, each step in the process must be reversible. For a step in a process to be reversible, the system must be in equilibrium throughout the step. That ideal cannot be accomplished in practice because no step can be taken without perturbing the system from equilibrium, but the ideal can be approached by making changes slowly.

Walls

by types of wall			
Types of transfers permitted **type of wall**	**type of transfer**		
	Matter	**Work**	**Heat**
permeable to matter	✓	✗	✗

permeable to energy but impermeable to matter	✗	✓	✓
adiabatic	✗	✓	✗
adynamic and impermeable to matter	✗	✗	✓
isolating	✗	✗	✗

A system is enclosed by walls that bound it and connect it to its surroundings. Often a wall restricts passage across it by some form of matter or energy, making the connection indirect. Sometimes a wall is no more than an imaginary two-dimensional closed surface through which the connection to the surroundings is direct.

A wall can be fixed (e.g. a constant volume reactor) or moveable (e.g. a piston). For example, in a reciprocating engine, a fixed wall means the piston is locked at its position; then, a constant volume process may occur. In that same engine, a piston may be unlocked and allowed to move in and out. Ideally, a wall may be declared adiabatic, diathermal, impermeable, permeable, or semi-permeable. Actual physical materials that provide walls with such idealized properties are not always readily available.

The system is delimited by walls or boundaries, either actual or notional, across which conserved (such as matter and energy) or unconserved (such as entropy) quantities can pass into and out of the system. The space outside the thermodynamic system is known as the *surroundings*, a *reservoir*, or the *environment*. The properties of the walls determine what transfers can occur. A wall that allows transfer of a quantity is said to be permeable to it, and a thermodynamic system is classified by the permeabilities of its several walls. A transfer between system and surroundings can arise by contact, such as conduction of heat, or by long-range forces such as an electric field in the surroundings.

A system with walls that prevent all transfers is said to be isolated. This is an idealized conception, because in practice some transfer is always possible, for example by gravitational forces. It is an axiom of thermodynamics that an isolated system eventually reaches internal thermodynamic equilibrium, when its state no longer changes with time.

The walls of a closed system allow transfer of energy as heat and as work, but not of matter, between it and its surroundings. The walls of an *open system* allow transfer both of matter and of energy. This scheme of definition of terms is not uniformly used, though it is convenient for some purposes. In particular, some writers use 'closed system' where 'isolated system' is here used.

Anything that passes across the boundary and effects a change in the contents of the system must be accounted for in an appropriate balance equation. The volume can be the region surrounding a single atom resonating energy, such as Max Planck defined in 1900; it can be a body of steam or air in a steam engine, such as Sadi Carnot defined in 1824. It could also be just one nuclide (i.e. a system of quarks) as hypothesized in quantum thermodynamics.

Surroundings

The system is the part of the universe being studied, while the *surroundings* is the remainder of the universe that lies outside the boundaries of the system. It is also known as the *environment*, and the *reservoir*. Depending on the type of system, it may interact with the system by exchanging mass, energy (including heat and work), momentum, electric charge, or other conserved properties. The environment is ignored in analysis of the system, except in regards to these interactions.

Closed System

In a closed system, no mass may be transferred in or out of the system boundaries. The system always contains the same amount of matter, but heat and work can be exchanged across the boundary of the system. Whether a system can exchange heat, work, or both is dependent on the property of its boundary.

- Adiabatic boundary – not allowing any heat exchange: A thermally isolated system

- Rigid boundary – not allowing exchange of work: A mechanically isolated system

One example is fluid being compressed by a piston in a cylinder. Another example of a closed system is a bomb calorimeter, a type of constant-volume calorimeter used in measuring the heat of combustion of a particular reaction. Electrical energy travels across the boundary to produce a spark between the electrodes and initiates combustion. Heat transfer occurs across the boundary after combustion but no mass transfer takes place either way.

Beginning with the first law of thermodynamics for an open system, this is expressed as:

$$\Delta U = Q - W + m_i \left(h + \frac{1}{2}v^2 + gz\right)_i - m_e \left(h + \frac{1}{2}v^2 + gz\right)_e$$

where U is internal energy, Q is the heat added to the system, W is the work done by the system, and since no mass is transferred in or out of the system, both expressions involving mass flow are zero and the first law of thermodynamics for a closed system is derived. The first law of thermodynamics for a closed system states that the increase of internal energy of the system equals the amount of heat added to the system minus the work done by the system. For infinitesimal changes the first law for closed systems is stated by:

$$dU = \delta Q - \delta W.$$

If the work is due to a volume expansion by dV at a pressure P then:

$$\delta W = PdV.$$

For a homogeneous system undergoing a reversible process, the second law of thermodynamics reads:

$$\delta Q = TdS$$

where T is the absolute temperature and S is the entropy of the system. With these relations the fundamental thermodynamic relation, used to compute changes in internal energy, is expressed as:

$$dU = TdS - PdV.$$

For a simple system, with only one type of particle (atom or molecule), a closed system amounts to a constant number of particles. However, for systems undergoing a chemical reaction, there may be all sorts of molecules being generated and destroyed by the reaction process. In this case, the fact that the system is closed is expressed by stating that the total number of each elemental atom is conserved, no matter what kind of molecule it may be a part of. Mathematically:

$$\sum_{j=1}^{m} a_{ij} N_j = b_i^0$$

where N_j is the number of j-type molecules, a_{ij} is the number of atoms of element i in molecule j and b_i^0 is the total number of atoms of element i in the system, which remains constant, since the system is closed. There is one such equation for each element in the system.

Isolated System

An isolated system is more restrictive than a closed system as it does not interact with its surroundings in any way. Mass and energy remains constant within the system, and no energy or mass transfer takes place across the boundary. As time passes in an isolated system, internal differences in the system tend to even out and pressures and temperatures tend to equalize, as do density differences. A system in which all equalizing processes have gone practically to completion is in a state of thermodynamic equilibrium.

Truly isolated physical systems do not exist in reality (except perhaps for the universe as a whole), because, for example, there is always gravity between a system with mass and masses elsewhere. However, real systems may behave nearly as an isolated system for finite (possibly very long) times. The concept of an isolated system can serve as a useful model approximating many real-world situations. It is an acceptable idealization used in constructing mathematical models of certain natural phenomena.

In the attempt to justify the postulate of entropy increase in the second law of thermodynamics, Boltzmann's H-theorem used equations, which assumed that a system (for example, a gas) was isolated. That is all the mechanical degrees of freedom could be specified, treating the walls simply as mirror boundary conditions. This inevitably led to Loschmidt's paradox. However, if the stochastic behavior of the molecules in actual walls is considered, along with the randomizing effect of the ambient, background thermal radiation, Boltzmann's assumption of molecular chaos can be justified.

The second law of thermodynamics for isolated systems states that the entropy of an isolated system not in equilibrium tends to increase over time, approaching maximum value at equilibrium. Overall, in an isolated system, the internal energy is constant and the entropy can never decrease. A *closed* system's entropy can decrease e.g. when heat is extracted from the system.

It is important to note that isolated systems are not equivalent to closed systems. Closed systems cannot exchange matter with the surroundings, but can exchange energy. Isolated systems can exchange neither matter nor energy with their surroundings, and as such are only theoretical and do not exist in reality (except, possibly, the entire universe).

It is worth noting that 'closed system' is often used in thermodynamics discussions when 'isolated system' would be correct - i.e. there is an assumption that energy does not enter or leave the system.

Selective Transfer of Matter

For a thermodynamic process, the precise physical properties of the walls and surroundings of the system are important, because they determine the possible processes.

An open system has one or several walls that allow transfer of matter. To account for the internal energy of the open system, this requires energy transfer terms in addition to those for heat and work. It also leads to the idea of the chemical potential.

A wall selectively permeable only to a pure substance can put the system in diffusive contact with a reservoir of that pure substance in the surroundings. Then a process is possible in which that pure substance is transferred between system and surroundings. Also, across that wall a contact equilibrium with respect to that substance is possible. By suitable thermodynamic operations, the pure substance reservoir can be dealt with as a closed system. Its internal energy and its entropy can be determined as functions of its temperature, pressure, and mole number.

A thermodynamic operation can render impermeable to matter all system walls other than the contact equilibrium wall for that substance. This allows the definition of an intensive state variable, with respect to a reference state of the surroundings, for that substance. The intensive variable is called the chemical potential; for component substance i it is usually denoted μ_i. The corresponding extensive variable can be the number of moles N_i of the component substance in the system.

For a contact equilibrium across a wall permeable to a substance, the chemical potentials of the substance must be same on either side of the wall. This is part of the nature of thermodynamic equilibrium, and may be regarded as related to the zeroth law of thermodynamics.

Open System

In an open system, matter may pass in and out of some segments of the system boundaries. There may be other segments of the system boundaries that pass heat or work but not matter. Respective account is kept of the transfers of energy across those and any other several boundary segments. In thermodynamic equilibrium, all flows have vanished.

Thermodynamic Process

Classical thermodynamics considers three main kinds of thermodynamic process: change in a system, cycles in a system, and flow processes.

Defined by change in a system, a thermodynamic process is a passage of a thermodynamic system from an initial to a final state of thermodynamic equilibrium. The initial and final states are the defining elements of the process. The actual course of the process is not the primary concern, and often is ignored. This is the customary default meaning of the term 'thermodynamic process'. In general, during the actual course of a thermodynamic process, the system passes through physical states which are not describable as thermodynamic states, because they are far from internal thermodynamic equilibrium. Such processes are useful for thermodynamic theory.

Defined by a cycle of transfers into and out of a system, a cyclic process is described by the quantities transferred in the several stages of the cycle, which recur unchangingly. The descriptions of the staged states of the system are not the primary concern. Cyclic processes were important conceptual devices in the early days of thermodynamical investigation, while the concept of the thermodynamic state variable was being developed.

Defined by flows through a system, a flow process is a steady state of flows into and out of a vessel with definite wall properties. The internal state of the vessel contents is not the primary concern. The quantities of primary concern describe the states of the inflow and the outflow materials, and, on the side, the transfers of heat, work, and kinetic and potential energies for the vessel. Flow processes are of interest in engineering.

Kinds of Process

Thermodynamic Process

Defined by change in a system, a thermodynamic process is a passage of a thermodynamic system from an initial to a final state of thermodynamic equilibrium. The initial and final states are the defining elements of the process. The actual course of the process is not the primary concern, and often is ignored. A state of thermodynamic equilibrium endures unchangingly unless it is interrupted by a thermodynamic operation that initiates a thermodynamic process. The equilibrium states are each respectively fully specified by a suitable set of thermodynamic state variables, that depend only on the current state of the system, not the path taken by the processes that produce that state. In general, during the actual course of a thermodynamic process, the system passes through physical states which are not describable as thermodynamic states, because they are far from internal thermodynamic equilibrium. Such a process may therefore be admitted for equilibrium thermodynamics, but not be admitted for non-equilibrium thermodynamics, which primarily aims to describe the continuous passage along the path, at definite rates of progress.

Though not so in general, it is, however, possible, that a process may take place slowly or smoothly enough to allow its description to be usefully approximated by a continuous path of equilibrium thermodynamic states. Then it may be approximately described by a process function that does depend on the path. Such a process may be idealized as a "quasi-static" process, which is infinitely slow, and which is really a theoretical exercise in differential geometry, as opposed to an actually possible physical process; in this idealized case, the calculation may be exact, though the process does not actually occur in nature. Such idealized processes are useful in the theory of thermodynamics.

Cyclic Process

Defined by a cycle of transfers into and out of a system, a cyclic process is described by the quantities transferred in the several stages of the cycle. The descriptions of the staged states of the system may be of little or even no interest. A cycle is a sequence of a small number of thermodynamic processes that indefinitely often repeatedly returns the system to its original state. For this, the staged states themselves are not necessarily described, because it is the transfers that are of interest. It is reasoned that if the cycle can be repeated indefinitely often, then it can be assumed that the states are recurrently unchanged. The condition of the system during the several staged processes may be of even less interest than is the precise nature of the recurrent states. If, however, the several staged processes are idealized and quasi-static, then the cycle is described by a path through a continuous progression of equilibrium states.

Flow Process

Defined by flows through a system, a flow process is a steady state of flow into and out of a vessel with definite wall properties. The internal state of the vessel contents is not the primary concern. The quantities of primary concern describe the states of the inflow and the outflow materials, and, on the side, the transfers of heat, work, and kinetic and potential energies for the vessel. The states of the inflow and outflow materials consist of their internal states, and of their kinetic and potential energies as whole bodies. Very often, the quantities that describe the internal states of the input and output materials are estimated on the assumption that they are bodies in their own states of internal thermodynamic equilibrium. Because rapid reactions are permitted, the thermodynamic treatment may be approximate, not exact.

A Cycle of Quasi-static Processes

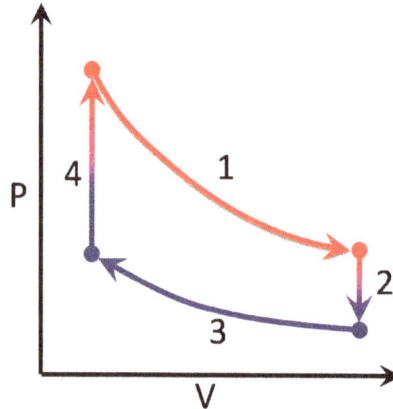

An example of a cycle of idealized thermodynamic processes which make up the Stirling cycle

A quasi-static thermodynamic process can be visualized by graphically plotting the path of idealized changes to the system's state variables. In the example, a cycle consisting of four quasi-static processes is shown. Each process has a well-defined start and end point in the pressure-volume state space. In this particular example, processes 1 and 3 are isothermal, whereas processes 2 and 4 are isochoric. The PV diagram is a particularly useful visualization of a quasi-static process, because the area under the curve of a process is the amount of work done by the system during that process. Thus work is considered to be a process variable, as its exact value depends on the particular path taken between the start and end points of the process. Similarly, heat may be transferred during a process, and it too is a process variable.

Conjugate Variable Processes

It is often useful to group processes into pairs, in which each variable held constant is one member of a conjugate pair.

Pressure - volume

The pressure-volume conjugate pair is concerned with the transfer of mechanical or dynamic energy as the result of work.

- An isobaric process occurs at constant pressure. An example would be to have a movable piston in a cylinder, so that the pressure inside the cylinder is always at atmospheric pressure, although it is separated from the atmosphere. In other words, the system is dynamically connected, by a movable boundary, to a constant-pressure reservoir.

- An isochoric process is one in which the volume is held constant, with the result that the mechanical PV work done by the system will be zero. On the other hand, work can be done isochorically on the system, for example by a shaft that drives a rotary paddle located inside the system. It follows that, for the simple system of one deformation variable, any heat energy transferred to the system externally will be absorbed as internal energy. An isochoric process is also known as an isometric process or an isovolumetric process. An example would be to place a closed tin can of material into a fire. To a first approximation, the can will not expand, and the only change will be that the contents gain internal energy, evidenced by increase in temperature and pressure. Mathematically, $\delta Q = dU$. The system is dynamically insulated, by a rigid boundary, from the environment.

Temperature - entropy

The temperature-entropy conjugate pair is concerned with the transfer of energy, especially for a closed system.

- An isothermal process occurs at a constant temperature. An example would be a closed system immersed in and thermally connected with a large constant-temperature bath. Energy gained by the system, through work done on it, is lost to the bath, so that its temperature remains constant.

- An adiabatic process is a process in which there is no matter or heat transfer, because a thermally insulating wall separates the system from its surroundings. For the process to be natural, either (a) work must be done on the system at a finite rate, so that the internal energy of the system increases; the entropy of the system increases even though it is thermally insulated; or (b) the system must do work on the surroundings, which then suffer increase of entropy, as well as gaining energy from the system.

- An isentropic process is customarily defined as an idealized quasi-static reversible adiabatic process, of transfer of energy as work. Otherwise, for a constant-entropy process, if work is done irreversibly, heat transfer is necessary, so that the process is not adiabatic, and an accurate artificial control mechanism is necessary; such is therefore not an ordinary natural thermodynamic process.

Chemical Potential - particle Number

The processes just above have assumed that the boundaries are also impermeable to particles. Otherwise, we may assume boundaries that are rigid, but are permeable to one or more types of particle. Similar considerations then hold for the chemical potential–particle number conjugate pair, which is concerned with the transfer of energy via this transfer of particles.

- In a *constant chemical potential process* the system is *particle-transfer connected*, by a particle-permeable boundary, to a constant-μ reservoir.

- The conjugate here is a constant particle number process. These are the processes outlined just above. There is no energy added or subtracted from the system by particle transfer. The system is *particle-transfer-insulated* from its environment by a boundary that is impermeable to particles, but permissive of transfers of energy as work or heat. These processes are the ones by which thermodynamic work and heat are defined, and for them, the system is said to be closed.

Thermodynamic Potentials

Any of the thermodynamic potentials may be held constant during a process. For example:

- An isenthalpic process introduces no change in enthalpy in the system.

Polytropic Processes

A polytropic process is a thermodynamic process that obeys the relation:

$$PV^n = C,$$

where P is the pressure, V is volume, n is any real number (the "polytropic index"), and C is a constant. This equation can be used to accurately characterize processes of certain systems, notably the compression or expansion of a gas, but in some cases, liquids and solids.

Processes Classified by the Second Law of Thermodynamics

According to Planck, one may think of three main classes of thermodynamic process: natural, fictively reversible, and impossible or unnatural.

Natural Process

Only natural processes occur in nature. For thermodynamics, a natural process is a transfer between systems that increases the sum of their entropies, and is irreversible. Natural processes may occur spontaneously, or may be triggered in a metastable or unstable system, as for example in the condensation of a supersaturated vapour.

Fictively Reversible Process

To describe the geometry of graphical surfaces that illustrate equilibrium relations between thermodynamic functions of state, one can fictively think of so-called "reversible processes". They are convenient theoretical objects that trace paths across graphical surfaces. They are called "processes" but do not describe naturally occurring processes, which are always irreversible. Because the points on the paths are points of thermodynamic equilibrium, it is customary to think of the "processes" described by the paths as fictively "reversible". Reversible processes are always quasistatic processes, but the converse is not always true.

Unnatural Process

Unnatural processes are logically conceivable but do not occur in nature. They would decrease the sum of the entropies if they occurred.

Quasistatic Process

A quasistatic process is an idealized or fictive model of a thermodynamic "process" considered in theoretical studies. It does not occur in physical reality. It may be imagined as happening infinitely slowly so that the system passes through a continuum of states that are infinitesimally close to equilibrium.

Thermodynamic Equilibrium

Thermodynamic equilibrium is an axiomatic concept of thermodynamics. It is an internal state of a single thermodynamic system, or a relation between several thermodynamic systems connected by more or less permeable or impermeable walls. In thermodynamic equilibrium there are no net macroscopic flows of matter or of energy, either within a system or between systems. In a system in its own state of internal thermodynamic equilibrium, no macroscopic change occurs. Systems in mutual thermodynamic equilibrium are simultaneously in mutual thermal, mechanical, chemical, and radiative equilibria. Systems can be in one kind of mutual equilibrium, though not in others. In thermodynamic equilibrium, all kinds of equilibrium hold at once and indefinitely, until disturbed by a thermodynamic operation. In a macroscopic equilibrium, almost or perfectly exactly balanced microscopic exchanges occur; this is the physical explanation of the notion of macroscopic equilibrium.

A thermodynamic system in its own state of internal thermodynamic equilibrium has a spatially uniform temperature. Its intensive properties, other than temperature, may be driven to spatial inhomogeneity by an unchanging long range force field imposed on it by its surroundings.

In non-equilibrium systems, by contrast, there are net flows of matter or energy. If such changes can be triggered to occur in a system in which they are not already occurring, it is said to be in a metastable equilibrium.

Though it is not a widely named law, it is an axiom of thermodynamics that there exist states of thermodynamic equilibrium. The second law of thermodynamics states that when a body of material starts from an equilibrium state, in which portions of it are held at different states by more or less permeable or impermeable partitions, and a thermodynamic operation removes or makes the partitions more permeable and it is isolated, then it spontaneously reaches its own new state of internal thermodynamic equilibrium, and this is accompanied by an increase in the sum of the entropies of the portions.

Overview

Classical thermodynamics deals with states of dynamic equilibrium. The state of a system at thermodynamic equilibrium is the one for which some thermodynamic potential is minimized, or for which the entropy (S) is maximized, for specified conditions. One such potential is the Helmholtz free energy (A), for a system with surroundings at controlled constant temperature and volume:

$$A = U - TS$$

Another potential, the Gibbs free energy (G), is minimized at thermodynamic equilibrium in a system with surroundings at controlled constant temperature and pressure:

$$G = U - TS + PV$$

where T denotes the absolute thermodynamic temperature, P the pressure, S the entropy, V the volume, and U the internal energy of the system.

Thermodynamic equilibrium is the unique stable stationary state that is approached or eventually reached as the system interacts with its surroundings over a long time. The above-mentioned potentials are mathematically constructed to be the thermodynamic quantities that are minimized under the particular conditions in the specified surroundings.

Conditions for Thermodynamic Equilibrium

- For a completely isolated system, S is maximum at thermodynamic equilibrium.

- For a system with controlled constant temperature and volume, A is minimum at thermodynamic equilibrium.

- For a system with controlled constant temperature and pressure, G is minimum at thermodynamic equilibrium.

The various types of equilibriums are achieved as follows:

- Two systems are in *thermal equilibrium* when their temperatures are the same.

- Two systems are in *mechanical equilibrium* when their pressures are the same.

- Two systems are in *diffusive equilibrium* when their chemical potentials are the same.

- All forces are balanced and there is no significant external driving force.

Relation of Exchange Equilibrium Between Systems

Often the surroundings of a thermodynamic system may also be regarded as another thermodynamic system. In this view, one may consider the system and its surroundings as two systems in mutual contact, with long-range forces also linking them. The enclosure of the system is the surface of contiguity or boundary between the two systems. In the thermodynamic formalism, that surface is regarded as having specific properties of permeability. For example, the surface of contiguity may be supposed to be permeable only to heat, allowing energy to transfer only as heat. Then the two systems are said to be in thermal equilibrium when the long-range forces are unchanging in time and the transfer of energy as heat between them has slowed and eventually stopped permanently; this is an example of a contact equilibrium. Other kinds of contact equilibrium are defined by other kinds of specific permeability. When two systems are in contact equilibrium with respect to a particular kind of permeability, they have common values of the intensive variable that belongs to that particular kind of permeability. Examples of such intensive variables are temperature, pressure, chemical potential.

A contact equilibrium may be regarded also as an exchange equilibrium. There is a zero balance of rate of transfer of some quantity between the two systems in contact equilibrium. For example, for a wall permeable only to heat, the rates of diffusion of internal energy as heat between the two systems are equal and opposite. An adiabatic wall between the two systems is 'permeable' only to energy transferred as work; at mechanical equilibrium the rates of transfer of energy as work between them are equal and opposite. If the wall is a simple wall, then the rates of transfer of volume across it are also equal and opposite; and the pressures on either side of it are equal. If the adiabatic wall is more complicated, with a sort of leverage, having an area-ratio, then the pressures of the two systems in exchange equilibrium are in the inverse ratio of the volume exchange ratio; this keeps the zero balance of rates of transfer as work.

A radiative exchange can occur between two otherwise separate systems. Radiative exchange equilibrium prevails when the two systems have the same temperature.

Thermodynamic State of Internal Equilibrium of a System

A collection of matter may be entirely isolated from its surroundings. If it has been left undisturbed for an indefinitely long time, classical thermodynamics postulates that it is in a state in which no changes occur within it, and there are no flows within it. This is a thermodynamic state of internal equilibrium. (This postulate is sometimes, but not often, called the "minus first" law of thermodynamics. One textbook calls it the "zeroth law", remarking that the authors think this more befitting that title than its more customary definition, which apparently was suggested by Fowler.)

Such states are a principal concern in what is known as classical or equilibrium thermodynamics, for they are the only states of the system that are regarded as well defined in that subject. A system in contact equilibrium with another system can by a thermodynamic operation be isolated, and upon the event of isolation, no change occurs in it. A system in a relation of contact equilibrium with another system may thus also be regarded as being in its own state of internal thermodynamic equilibrium.

Multiple Contact Equilibrium

The thermodynamic formalism allows that a system may have contact with several other systems at once, which may or may not also have mutual contact, the contacts having respectively different permeabilities. If these systems are all jointly isolated from the rest of the world, those of them that are in contact then reach respective contact equilibria with one another.

If several systems are free of adiabatic walls between each other, but are jointly isolated from the rest of the world, then they reach a state of multiple contact equilibrium, and they have a common temperature, a total internal energy, and a total entropy. Amongst intensive variables, this is a unique property of temperature. It holds even in the presence of long-range forces. (That is, there is no "force" that can maintain temperature discrepancies.) For example, in a system in thermodynamic equilibrium in a vertical gravitational field, the pressure on the top wall is less than that on the bottom wall, but the temperature is the same everywhere.

A thermodynamic operation may occur as an event restricted to the walls that are within the surroundings, directly affecting neither the walls of contact of the system of interest with its surroundings, nor its interior, and occurring within a definitely limited time. For example, an immovable adiabatic wall may be placed or removed within the surroundings. Consequent upon such an operation restricted to the surroundings, the system may be for a time driven away from its own initial internal state of thermodynamic equilibrium. Then, according to the second law of thermodynamics, the whole undergoes changes and eventually reaches a new and final equilibrium with the surroundings. Following Planck, this consequent train of events is called a natural thermodynamic process. It is allowed in equilibrium thermodynamics just because the initial and final states are of thermodynamic equilibrium, even though during the process there is transient departure from thermodynamic equilibrium, when neither the system nor its surroundings are in well defined states of internal equilibrium. A natural process proceeds at a finite rate for the main part of its course. It is thereby radically different from a fictive quasi-static 'process' that proceeds infinitely slowly throughout its course, and is fictively 'reversible'. Classical thermodynamics allows that even though a process may take a very long time to settle to thermodynamic equilibrium, if the main part of its course is at a finite rate, then it is considered to be natural, and to be subject to the second law of thermodynamics, and thereby irreversible. Engineered machines and artificial devices and manipulations are permitted within the surroundings. The allowance of such operations and devices in the surroundings but not in the system is the reason why Kelvin in one of his statements of the second law of thermodynamics spoke of "inanimate" agency; a system in thermodynamic equilibrium is inanimate.

Otherwise, a thermodynamic operation may directly affect a wall of the system.

It is often convenient to suppose that some of the surrounding subsystems are so much larger than the system that the process can affect the intensive variables only of the surrounding subsystems, and they are then called reservoirs for relevant intensive variables.

Local and Global Equilibrium

It is useful to distinguish between global and local thermodynamic equilibrium. In thermodynamics, exchanges within a system and between the system and the outside are controlled by intensive parameters. As an example, temperature controls heat exchanges. *Global thermodynamic equilibrium* (GTE) means that those intensive parameters are homogeneous throughout the whole system, while *local thermodynamic equilibrium* (LTE) means that those intensive parameters are varying in space and time, but are varying so slowly that, for any point, one can assume thermodynamic equilibrium in some neighborhood about that point.

If the description of the system requires variations in the intensive parameters that are too large, the very assumptions upon which the definitions of these intensive parameters are based will break down, and the system will be in neither global nor local equilibrium. For example, it takes a certain number of collisions for a particle to equilibrate to its surroundings. If the average distance it has moved during these collisions removes it from the neighborhood it is equilibrating to, it will never equilibrate, and there will be no LTE. Temperature is, by definition, proportional to the average internal energy of an equilibrated neighborhood. Since there is no equilibrated neighborhood, the concept of temperature breaks down, and the temperature becomes undefined.

It is important to note that this local equilibrium may apply only to a certain subset of particles in the system. For example, LTE is usually applied only to massive particles. In a radiating gas, the photons being emitted and absorbed by the gas need not be in thermodynamic equilibrium with each other or with the massive particles of the gas in order for LTE to exist. In some cases, it is not considered necessary for free electrons to be in equilibrium with the much more massive atoms or molecules for LTE to exist.

As an example, LTE will exist in a glass of water that contains a melting ice cube. The temperature inside the glass can be defined at any point, but it is colder near the ice cube than far away from it. If energies of the molecules located near a given point are observed, they will be distributed according to the Maxwell–Boltzmann distribution for a certain temperature. If the energies of the molecules located near another point are observed, they will be distributed according to the Maxwell–Boltzmann distribution for another temperature.

Local thermodynamic equilibrium does not require either local or global stationarity. In other words, each small locality need not have a constant temperature. However, it does require that each small locality change slowly enough to practically sustain its local Maxwell–Boltzmann distribution of molecular velocities. A global non-equilibrium state can be stably stationary only if it is maintained by exchanges between the system and the outside. For example, a globally-stable stationary state could be maintained inside the glass of water by continuously adding finely powdered ice into it in order to compensate for the melting, and continuously draining off the meltwater. Natural transport phenomena may lead a system from local to global thermodynamic equilibrium. Going back to our example, the diffusion of heat will lead our glass of water toward global thermodynamic equilibrium, a state in which the temperature of the glass is completely homogeneous.

Reservations

Careful and well informed writers about thermodynamics, in their accounts of thermodynamic equilibrium, often enough make provisos or reservations to their statements. Some writers leave such reservations merely implied or more or less unstated.

For example, one widely cited writer, H. B. Callen writes in this context: "In actuality, few systems are in absolute and true equilibrium." He refers to radioactive processes and remarks that they may take "cosmic times to complete, [and] generally can be ignored". He adds "In practice, the criterion for equilibrium is circular. *Operationally, a system is in an equilibrium state if its properties are consistently described by thermodynamic theory!*"

J.A. Beattie and I. Oppenheim write: "Insistence on a strict interpretation of the definition of equilibrium would rule out the application of thermodynamics to practically all states of real systems."

Another author, cited by Callen as giving a "scholarly and rigorous treatment", and cited by Adkins as having written a "classic text", A.B. Pippard writes in that text: "Given long enough a supercooled vapour will eventually condense, …. The time involved may be so enormous, however, perhaps 10 years or more, …. For most purposes, provided the rapid change is not artificially stimulated, the systems may be regarded as being in equilibrium."

Another author, A. Münster, writes in this context. He observes that thermonuclear processes often occur so slowly that they can be ignored in thermodynamics. He comments: "The concept 'absolute equilibrium' or 'equilibrium with respect to all imaginable processes', has therefore, no physical significance." He therefore states that: "... we can consider an equilibrium only with respect to specified processes and defined experimental conditions."

According to L. Tisza: "... in the discussion of phenomena near absolute zero. The absolute predictions of the classical theory become particularly vague because the occurrence of frozen-in nonequilibrium states is very common."

Definitions

The most general kind of thermodynamic equilibrium of a system is through contact with the surroundings that allows simultaneous passages of all chemical substances and all kinds of energy. A system in thermodynamic equilibrium may move with uniform acceleration through space but must not change its shape or size while doing so; thus it is defined by a rigid volume in space. It may lie within external fields of force, determined by external factors of far greater extent than the system itself, so that events within the system cannot in an appreciable amount affect the external fields of force. The system can be in thermodynamic equilibrium only if the external force fields are uniform, and are determining its uniform acceleration, or if it lies in a non-uniform force field but is held stationary there by local forces, such as mechanical pressures, on its surface.

Thermodynamic equilibrium is a primitive notion of the theory of thermodynamics. According to P.M. Morse: "It should be emphasized that the fact that there are thermodynamic states, ..., and the fact that there are thermodynamic variables which are uniquely specified by the equilibrium state ... are *not* conclusions deduced logically from some philosophical first principles. They are conclusions ineluctably drawn from more than two centuries of experiments." This means that thermodynamic equilibrium is not to be defined solely in terms of other theoretical concepts of thermodynamics. M. Bailyn proposes a fundamental law of thermodynamics that defines and postulates the existence of states of thermodynamic equilibrium.

Textbook definitions of thermodynamic equilibrium are often stated carefully, with some reservation or other.

For example, A. Münster writes: "An isolated system is in thermodynamic equilibrium when, in the system, no changes of state are occurring at a measurable rate." There are two reservations stated here; the system is isolated; any changes of state are immeasurably slow. He discusses the second proviso by giving an account of a mixture oxygen and hydrogen at room temperature in the absence of a catalyst. Münster points out that a thermodynamic equilibrium state is described by fewer macroscopic variables than is any other state of a given system. This is partly, but not entirely, because all flows within and through the system are zero.

R. Haase's presentation of thermodynamics does not start with a restriction to thermodynamic equilibrium because he intends to allow for non-equilibrium thermodynamics. He considers an arbitrary system with time invariant properties. He tests it for thermodynamic equilibrium by cutting it off from all external influences, except external force fields. If after insulation, nothing changes, he says that the system was in *equilibrium*.

In a section headed "Thermodynamic equilibrium", H.B. Callen defines equilibrium states in a paragraph. He points out that they "are determined by intrinsic factors" within the system. They are "terminal states", towards which the systems evolve, over time, which may occur with "glacial slowness". This statement does not explicitly say that for thermodynamic equilibrium, the system must be isolated; Callen does not spell out what he means by the words "intrinsic factors".

Another textbook writer, C.J. Adkins, explicitly allows thermodynamic equilibrium to occur in a system which is not isolated. His system is, however, closed with respect to transfer of matter. He writes: "In general, the approach to thermodynamic equilibrium will involve both thermal and work-like interactions with the surroundings." He distinguishes such thermodynamic equilibrium from thermal equilibrium, in which only thermal contact is mediating transfer of energy.

Another textbook author, J.R. Partington, writes: "(i) *An equilibrium state is one which is independent of time.*" But, referring to systems "which are only apparently in equilibrium", he adds : "Such systems are in states of "false equilibrium."" Partington's statement does not explicitly state that the equilibrium refers to an isolated system. Like Münster, Partington also refers to the mixture of oxygen and hydrogen. He adds a proviso that "In a true equilibrium state, the smallest change of any external condition which influences the state will produce a small change of state ..." This proviso means that thermodynamic equilibrium must be stable against small perturbations; this requirement is essential for the strict meaning of thermodynamic equilibrium.

A student textbook by F.H. Crawford has a section headed "Thermodynamic Equilibrium". It distinguishes several drivers of flows, and then says: "These are examples of the apparently universal tendency of isolated systems toward a state of complete mechanical, thermal, chemical, and electrical—or, in a single word, *thermodynamic—equilibrium.*"

A monograph on classical thermodynamics by H.A. Buchdahl considers the "equilibrium of a thermodynamic system", without actually writing the phrase "thermodynamic equilibrium". Referring to systems closed to exchange of matter, Buchdahl writes: "If a system is in a terminal condition which is properly static, it will be said to be in *equilibrium.*" Buchdahl's monograph also discusses amorphous glass, for the purposes of thermodynamic description. It states: "More precisely, the glass may be regarded as being *in equilibrium* so long as experimental tests show that 'slow' transitions are in effect reversible." It is not customary to make this proviso part of the definition of thermodynamic equilibrium, but the converse is usually assumed: that if a body in thermodynamic equilibrium is subject to a sufficiently slow process, that process may be considered to be sufficiently nearly reversible, and the body remains sufficiently nearly in thermodynamic equilibrium during the process.

A. Münster carefully extends his definition of thermodynamic equilibrium for isolated systems by introducing a concept of *contact equilibrium*. This specifies particular processes that are allowed when considering thermodynamic equilibrium for non-isolated systems, with special concern for open systems, which may gain or lose matter from or to their surroundings. A contact equilibrium is between the system of interest and a system in the surroundings, brought into contact with the system of interest, the contact being through a special kind of wall; for the rest, the whole joint system is isolated. Walls of this special kind were also considered by C. Carathéodory, and are mentioned by other writers also. They are selectively permeable. They may be permeable only to mechanical work, or only to heat, or only to some particular chemical substance. Each contact equilibrium defines an intensive parameter; for example, a wall permeable only to heat defines an empirical temperature. A contact equilibrium can exist for each chemical constituent of the system of interest. In a contact equilibrium, despite the possible exchange through the selectively permeable wall, the system of interest is changeless, as if it were in isolated thermodynamic equilibrium. This scheme follows the general rule that "... we can consider an equilibrium only with respect to specified processes and defined experimental conditions." Thermodynamic equilibrium for an open system means that, with respect to every relevant kind of selectively permeable wall, contact equilibrium exists when the respective intensive parameters of the system and surroundings are equal. This definition does not consider the most general kind of thermodynamic equilibrium, which is through unselective contacts. This definition does not simply state that no current of matter or energy exists in the interior or at the boundaries; but it is compatible with the following definition, which does so state.

M. Zemansky also distinguishes mechanical, chemical, and thermal equilibrium. He then writes: "When the conditions for all three types of equilibrium are satisfied, the system is said to be in a state of thermodynamic equilibrium".

P.M. Morse writes that thermodynamics is concerned with "*states of thermodynamic equilibrium*". He also uses the phrase "thermal equilibrium" while discussing transfer of energy as heat between a body and a heat reservoir in its surroundings, though not explicitly defining a special term 'thermal equilibrium'.

J.R. Waldram writes of "a definite thermodynamic state". He defines the term "thermal equilibrium" for a system "when its observables have ceased to change over time". But shortly below that definition he writes of a piece of glass that has not yet reached its "*full* thermodynamic equilibrium state".

Considering equilibrium states, M. Bailyn writes: "Each intensive variable has its own type of equilibrium." He then defines thermal equilibrium, mechanical equilibrium, and material equilibrium. Accordingly, he writes: "If all the intensive variables become uniform, *thermodynamic equilibrium* is said to exist." He is not here considering the presence of an external force field.

J.G. Kirkwood and I. Oppenheim define thermodynamic equilibrium as follows: "A system is in a state of *thermodynamic equilibrium* if, during the time period allotted for experimentation, (a) its intensive properties are independent of time and (b) no current of matter or energy exists in its interior or at its boundaries with the surroundings." It is evident that they are not restricting the definition to isolated or to closed systems. They do not discuss the possibility of changes that occur with "glacial slowness", and proceed beyond the time period allotted for experimentation. They note that for two systems in contact, there exists a small subclass of intensive properties such that if all those of that small subclass are respectively equal, then all respective intensive properties are equal. States of thermodynamic equilibrium may be defined by this subclass, provided some other conditions are satisfied.

Characteristics of a State of Internal Thermodynamic Equilibrium

Homogeneity in the Absence of External Forces

A thermodynamic system consisting of a single phase in the absence of external forces, in its own internal thermodynamic equilibrium, is homogeneous. This means that the material in any small volume element of the system can be interchanged with the material of any other geometrically congruent volume element of the system, and the effect is to leave the system thermodynamically unchanged. In general, a strong external force field makes a system of a single phase in its own internal thermodynamic equilibrium inhomogeneous with respect to some intensive variables. For example, a relatively dense component of a mixture can be concentrated by centrifugation.

Uniform Temperature

Such equilibrium inhomogeneity, induced by external forces, does not occur for the intensive variable temperature. According to E.A. Guggenheim, "The most important conception of thermodynamics is temperature." Planck introduces his treatise with a brief account of heat and temperature and thermal equilibrium, and then announces: "In the following we shall deal chiefly with homogeneous, isotropic bodies of any form, possessing throughout their substance the same temperature and density, and subject to a uniform pressure acting everywhere perpendicular to the surface." As did Carathéodory, Planck was setting aside surface effects and external fields and anisotropic crystals. Though referring to temperature, Planck did not there explicitly refer to the concept of thermodynamic equilibrium. In contrast, Carathéodory's scheme of presentation of classical thermodynamics for closed systems postulates the concept of an "equilibrium state" following Gibbs (Gibbs speaks routinely of a "thermodynamic state"), though not explicitly using the phrase 'thermodynamic equilibrium', nor explicitly postulating the existence of a temperature to define it.

The temperature within a system in thermodynamic equilibrium is uniform in space as well as in time. In a system in its own state of internal thermodynamic equilibrium, there are no net internal macroscopic flows. In particular, this means that all local parts of the system are in mutual radiative exchange equilibrium. This means that the temperature of the system is spatially uniform. This is so in all cases, including those of non-uniform external force fields. For an externally imposed gravitational field, this may be proved in macroscopic thermodynamic terms, by the calculus of variations, using the method of Langrangian multipliers. Considerations of kinetic theory or statistical mechanics also support this statement.

In order that a system may be in its own internal state of thermodynamic equilibrium, it is of course necessary, but not sufficient, that it be in its own internal state of thermal equilibrium; it is possible for a system to reach internal mechanical equilibrium before it reaches internal thermal equilibrium.

Number of Real Variables Needed for Specification

In his exposition of his scheme of closed system equilibrium thermodynamics, C. Carathéodory initially postulates that experiment reveals that a definite number of real variables define the states that are the points of the manifold of equilibria. In the words of Prigogine and Defay (1945): "It is a matter of experience that when we have specified a certain number of macroscopic properties of a system, then all the other properties are fixed." As noted above, according to A. Münster, the number of variables needed to define a thermodynamic equilibrium is the least for any state of a given isolated system. As noted above, J.G. Kirkwood and I. Oppenheim point out that a state of thermodynamic equilibrium may be defined by a special subclass of intensive variables, with a definite number of members in that subclass.

If the thermodynamic equilibrium lies in an external force field, it is only the temperature that can in general be expected to be spatially uniform. Intensive variables other than temperature will in general be non-uniform if the external force field is non-zero. In such a case, in general, additional variables are needed to describe the spatial non-uniformity.

Stability Against Small Perturbations

As noted above, J.R. Partington points out that a state of thermodynamic equilibrium is stable against small transient perturbations. Without this condition, in general, experiments intended to study systems in thermodynamic equilibrium are in severe difficulties.

Approach to Thermodynamic Equilibrium Within an Isolated System

When a body of material starts from a non-equilibrium state of inhomogeneity or chemical non-equilibrium, and is then isolated, it spontaneously evolves towards its own internal state of thermodynamic equilibrium. It is not necessary that all aspects of internal thermodynamic equilibrium be reached simultaneously; some can be established before others. For example, in many cases of such evolution, internal mechanical equilibrium is established much more rapidly than the other aspects of the eventual thermodynamic equilibrium. Another example is that, in many cases of such evolution, thermal equilibrium is reached much more rapidly than chemical equilibrium.

Fluctuations within an Isolated System in Its Own Internal Thermodynamic Equilibrium

In an isolated system, thermodynamic equilibrium by definition persists over an indefinitely long time. In classical physics it is often convenient to ignore the effects of measurement and this is assumed in the present account.

To consider the notion of fluctuations in an isolated thermodynamic system, a convenient example is a system specified by its extensive state variables, internal energy, volume, and mass composition. By definition they are time-invariant. By definition, they combine with time-invariant nominal values of their conjugate intensive functions of state, inverse temperature, pressure divided by temperature, and the chemical potentials divided by temperature, so as to exactly obey the laws of thermodynamics. But the laws of thermodynamics, combined with the values of the specifying extensive variables of state, are not sufficient to provide knowledge of those nominal values.

It may be admitted that on repeated measurement of those conjugate intensive functions of state, they are found to have slightly different values from time to time. Such variability is regarded as due to internal fluctuations. The different measured values average to their nominal values.

If the system is truly macroscopic as postulated by classical thermodynamics, then the fluctuations are too small to detect macroscopically. This is called the thermodynamic limit. In effect, the molecular nature of matter and the quantal nature of momentum transfer have vanished from sight, too small to see. According to Buchdahl: "... there is no place within the strictly phenomenological theory for the idea of fluctuations about equilibrium."

If the system is repeatedly subdivided, eventually a system is produced that is small enough to exhibit obvious fluctuations. This is a mesoscopic level of investigation. The fluctuations are then directly dependent on the natures of the various walls of the system. The precise choice of independent state variables is then important. At this stage, statistical features of the laws of thermodynamics become apparent.

If the mesoscopic system is further repeatedly divided, eventually a microscopic system is produced. Then the molecular character of matter and the quantal nature of momentum transfer become important in the processes of fluctuation. One has left the realm of classical or macroscopic thermodynamics, and one needs quantum statistical mechanics. The fluctuations can become relatively dominant, and questions of measurement become important.

The statement that 'the system is its own internal thermodynamic equilibrium' may be taken to mean that 'indefinitely many such measurements have been taken from time to time, with no trend in time in the various measured values'. Thus the statement, that 'a system is in its own internal thermodynamic equilibrium, with stated nominal values of its functions of state conjugate to its specifying state variables', is far far more informative than a statement that 'a set of single simultaneous measurements of those functions of state have those same values'. This is because the single measurements might have been made during a slight fluctuation, away from another set of nominal values of those conjugate intensive functions of state, that is due to unknown and different constitutive properties. A single measurement cannot tell whether that might be so, unless there is also knowledge of the nominal values that belong to the equilibrium state.

Thermal Equilibrium

An explicit distinction between 'thermal equilibrium' and 'thermodynamic equilibrium' is made by B. C. Eu. He considers two systems in thermal contact, one a thermometer, the other a system in which there are occurring several irreversible processes, entailing non-zero fluxes; the two systems are separated by a wall permeable only to heat. He considers the case in which, over the time scale of interest, it happens that both the thermometer reading and the irreversible processes are steady. Then there is thermal equilibrium without thermodynamic equilibrium. Eu proposes consequently that the zeroth law of thermodynamics can be considered to apply even when thermodynamic equilibrium is not present; also he proposes that if changes are occurring so fast that a steady temperature cannot be defined, then "it is no longer possible to describe the process by means of a thermodynamic formalism. In other words, thermodynamics has no meaning for such a process." This illustrates the importance for thermodynamics of the concept of temperature.

Thermal equilibrium is achieved when two systems in thermal contact with each other cease to have a net exchange of energy. It follows that if two systems are in thermal equilibrium, then their temperatures are the same.

Thermal equilibrium occurs when a system's macroscopic thermal observables have ceased to change with time. For example, an ideal gas whose distribution function has stabilised to a specific Maxwell–Boltzmann distribution would be in thermal equilibrium. This outcome allows a single temperature and pressure to be attributed to the whole system. For an isolated body, it is quite possible for mechanical equilibrium to be reached before thermal equilibrium is reached, but eventually, all aspects of equilibrium, including thermal equilibrium, are necessary for thermodynamic equilibrium.

Non-equilibrium

A system's internal state of thermodynamic equilibrium should be distinguished from a "stationary state" in which thermodynamic parameters are unchanging in time but the system is not isolated, so that there are, into and out of the system, non-zero macroscopic fluxes which are constant in time.

Non-equilibrium thermodynamics is a branch of thermodynamics that deals with systems that are not in thermodynamic equilibrium. Most systems found in nature are not in thermodynamic equilibrium because they are changing or can be triggered to change over time, and are continuously and discontinuously subject to flux of matter and energy to and from other systems. The thermodynamic study of non-equilibrium systems requires more general concepts than are dealt with by equilibrium thermodynamics. Many natural systems still today remain beyond the scope of currently known macroscopic thermodynamic methods.

References

- Bailyn, M. (1994). A Survey of Thermodynamics, American Institute of Physics Press, New York, ISBN 0-88318-797-3, p. 20.

- Eu, B.C. (2002). Generalized Thermodynamics. The Thermodynamics of Irreversible Processes and Generalized Hydrodynamics, Kluwer Academic Publishers, Dordrecht, ISBN 1-4020-0788-4.

- Tschoegl, N.W. (2000). Fundamentals of Equilibrium and Steady-State Thermodynamics, Elsevier, Amsterdam, ISBN 0-444-50426-5, p. 5.

- Callen, H.B. (1960/1985). Thermodynamics and an Introduction to Thermostatistics, (1st edition 1960) 2nd edition 1985, Wiley, New York, ISBN 0-471-86256-8, p. 17.

- I.M.Kolesnikov; V.A.Vinokurov; S.I.Kolesnikov (2001). Thermodynamics of Spontaneous and Non-Spontaneous Processes. Nova science Publishers. p. 136. ISBN 1-56072-904-X.

- Bailyn, M. (1994). A Survey of Thermodynamics, American Institute of Physics Press, New York, ISBN 0-88318-797-3, pp. 19–23.

- Callen, H.B. (1960/1985). Thermodynamics and an Introduction to Thermostatistics, (1st edition 1960) 2nd edition 1985, Wiley, New York, ISBN 0-471-86256-8.

- Tschoegl, N.W. (2000). Fundamentals of Equilibrium and Steady-State Thermodynamics, Elsevier, Amsterdam, ISBN 0-444-50426-5, p. 21.

- Bailyn, M. (1994). A Survey of Thermodynamics, American Institute of Physics Press, New York, ISBN 0-88318-797-3.

- Beattie, J.A., Oppenheim, I. (1979). Principles of Thermodynamics, Elsevier Scientific Publishing, Amsterdam, ISBN 0-444-41806-7.

- Callen, H.B. (1960/1985). Thermodynamics and an Introduction to Thermostatistics, (1st edition 1960) 2nd edition 1985, Wiley, New York, ISBN 0-471-86256-8.

- de Groot, S.R., Mazur, P. (1962). Non-equilibrium Thermodynamics, North-Holland, Amsterdam. Reprinted (1984), Dover Publications Inc., New York, ISBN 0486647412.

- Eu, B.C. (2002). Generalized Thermodynamics. The Thermodynamics of Irreversible Processes and Generalized Hydrodynamics, Kluwer Academic Publishers, Dordrecht, ISBN 1-4020-0788-4.

- Griem, H.R. (2005). Principles of Plasma Spectroscopy (Cambridge Monographs on Plasma Physics), Cambridge University Press, New York ISBN 0-521-61941-6.

- Waldram, J.R. (1985). The Theory of Thermodynamics, Cambridge University Press, Cambridge UK, ISBN 0-521-24575-3.

8

Diverse Aspects of Thermal Engineering

Thermal engineering has various features; some of these aspects are thermal conduction, thermal radiation, thermal grease, thermal mass, heat and heat capacity. Thermal conduction is the transmitting of heat by atoms and the movement of electrons within a body. The diverse aspects of thermal engineering in the current scenario have been thoroughly discussed in this chapter.

Thermal Conduction

Thermal conduction is the transfer of heat (internal energy) by microscopic collisions of particles and movement of electrons within a body. The microscopically colliding objects, that include molecules, atoms, and electrons, transfer disorganized microscopic kinetic and potential energy, jointly known as internal energy. Conduction takes place in all phases of matter, such as solids, liquids, gases and plasmas. The rate at which energy is conducted as heat between two bodies is a function of the temperature difference (temperature gradient) between the two bodies and the properties of the conductive medium through which the heat is transferred. Thermal conduction was originally called diffusion.

Heat spontaneously flows from a hotter to a colder body. For example, heat is conducted from the hotplate of an electric stove to the bottom of a saucepan in contact with it. In the absence of an external driving energy source to the contrary, within a body or between bodies, temperature differences decay over time, and thermal equilibrium is approached, temperature becoming more uniform.

In conduction, the heat flow is within and through the body itself. In contrast, in heat transfer by thermal radiation, the transfer is often between bodies, which may be separated spatially. Also possible is transfer of heat by a combination of conduction and thermal radiation. In convection, internal energy is carried between bodies by a moving material carrier. In solids, conduction is mediated by the combination of vibrations and collisions of molecules, of propagation and collisions of phonons, and of diffusion and collisions of free electrons. In gases and liquids, conduction is due to the collisions and diffusion of molecules during their random motion. Photons in this context do not collide with one another, and so heat transport by electromagnetic radiation is conceptually distinct from heat conduction by microscopic diffusion and collisions of material particles and phonons. But the distinction is often not easily observed, unless the material is semi-transparent.

In the engineering sciences, heat transfer includes the processes of thermal radiation, convection, and sometimes mass transfer. Usually, more than one of these processes occurs in a given situation. The conventional symbol for the material property, thermal conductivity, is .

Overview

On a microscopic scale, conduction occurs within a body considered as being stationary; this means that the kinetic and potential energies of the bulk motion of the body are separately accounted for. Internal energy diffuses as rapidly moving or vibrating atoms and molecules interact with neighboring particles, transferring some of their microscopic kinetic and potential energies, these quantities being defined relative to the bulk of the body considered as being stationary. Heat is transferred by conduction when adjacent atoms or molecules collide, or as several electrons move backwards and forwards from atom to atom in a disorganized way so as not to form a macroscopic electric current, or as phonons collide and scatter. Conduction is the most significant means of heat transfer within a solid or between solid objects in thermal contact. Conduction is greater in solids because the network of relatively close fixed spatial relationships between atoms helps to transfer energy between them by vibration.

Fluids (and especially gases) are less conductive. This is due to the large distance between atoms in a gas: fewer collisions between atoms means less conduction. At low densities, the conductivity of gases can be derived from kinetic theory using rigid, non-interacting spheres and refined using Chapman-Enskog kinetic theory of gases. The conductivity of gases increases with temperature. Conductivity increases with increasing pressure from vacuum up to a critical point that the density of the gas is such that molecules of the gas may be expected to collide with each other before they transfer heat from one surface to another. After this point, conductivity increases only slightly with increasing pressure and density.

Thermal contact conductance is the study of heat conduction between solid bodies in contact. A temperature drop is often observed at the interface between the two surfaces in contact. This phenomenon is said to be a result of a thermal contact resistance existing between the contacting surfaces. Interfacial thermal resistance is a measure of an interface's resistance to thermal flow. This thermal resistance differs from contact resistance, as it exists even at atomically perfect interfaces. Understanding the thermal resistance at the interface between two materials is of primary significance in the study of its thermal properties. Interfaces often contribute significantly to the observed properties of the materials.

The inter-molecular transfer of energy could be primarily by elastic impact, as in fluids, or by free electron diffusion, as in metals, or phonon vibration, as in insulators. In insulators, the heat flux is carried almost entirely by phonon vibrations.

Metals (e.g., copper, platinum, gold, etc.) are usually good conductors of thermal energy. This is due to the way that metals bond chemically: metallic bonds (as opposed to covalent or ionic bonds) have free-moving electrons that transfer thermal energy rapidly through the metal. The *electron fluid* of a conductive metallic solid conducts most of the heat flux through the solid. Phonon flux is still present, but carries less of the energy. Electrons also conduct electric current through conductive solids, and the thermal and electrical conductivities of most metals have about the same ratio. A good electrical conductor, such as copper, also conducts heat well. Thermoelectricity is caused by the interaction of heat flux and electric current. Heat conduction within a solid is directly analogous to diffusion of particles within a fluid, in the situation where there are no fluid currents.

To quantify the ease with which a particular medium conducts, engineers employ the thermal conductivity, also known as the conductivity constant or conduction coefficient, k. In thermal conductivity, k is defined as "the quantity of heat, Q, transmitted in time (t) through a thickness (L), in a direction normal to a surface of area (A), due to a temperature difference (ΔT) [...]". Thermal conductivity is a material *property* that is primarily dependent on the medium's phase, temperature, density, and molecular bonding. Thermal effusivity is a quantity derived from conductivity, which is a measure of its ability to exchange thermal energy with its surroundings.

Steady-state Conduction

Steady state conduction is the form of conduction that happens when the temperature difference(s) driving the conduction are constant, so that (after an equilibration time), the spatial distribution of temperatures (temperature field) in the conducting object does not change any further. Thus, all partial derivatives of temperature *with respect to space* may either be zero or have nonzero values, but all derivatives of temperature at any point *with respect to time* are uniformly zero. In steady state conduction, the amount of heat entering any region of an object is equal to amount of heat coming out (if this were not so, the temperature would be rising or falling, as thermal energy was tapped or trapped in a region).

For example, a bar may be cold at one end and hot at the other, but after a state of steady state conduction is reached, the spatial gradient of temperatures along the bar does not change any further, as time proceeds. Instead, the temperature at any given section of the rod remains constant, and this temperature varies linearly in space, along the direction of heat transfer.

In steady state conduction, all the laws of direct current electrical conduction can be applied to "heat currents". In such cases, it is possible to take "thermal resistances" as the analog to electrical resistances. In such cases, temperature plays the role of voltage, and heat transferred per unit time (heat power) is the analog of electric current. Steady state systems can be modelled by networks of such thermal resistances in series and in parallel, in exact analogy to electrical networks of resistors.

Transient Conduction

In general, during any period in which temperatures change *in time* at any place within an object, the mode of thermal energy flow is termed *transient conduction*. Another term is "non steady-state" conduction, referring to time-dependence of temperature fields in an object. Non-steady-state situations appear after an imposed change in temperature at a boundary of an object. They may also occur with temperature changes inside an object, as a result of a new source or sink of heat suddenly introduced within an object, causing temperatures near the source or sink to change in time.

When a new perturbation of temperature of this type happens, temperatures within the system change in time toward a new equilibrium with the new conditions, provided that these do not change. After equilibrium, heat flow into the system once again equals the heat flow out, and temperatures at each point inside the system no longer change. Once this happens, transient conduction is ended, although steady-state conduction may continue if heat flow continues.

If changes in external temperatures or internal heat generation changes are too rapid for the equilibrium of temperatures in space to take place, then the system never reaches a state of unchanging temperature distribution in time, and the system remains in a transient state.

An example of a new source of heat "turning on" within an object, causing transient conduction, is an engine starting in an automobile. In this case, the transient thermal conduction phase for the entire machine is over, and the steady state phase appears, as soon as the engine reaches steady-state operating temperature. In this state of steady-state equilibrium, temperatures vary greatly from the engine cylinders to other parts of the automobile, but at no point in space within the automobile does temperature increase or decrease. After establishing this state, the transient conduction phase of heat transfer is over.

New external conditions also cause this process: for example the copper bar in the example steady-state conduction experiences transient conduction as soon as one end is subjected to a different temperature from the other. Over time, the field of temperatures inside the bar reach a new steady-state, in which a constant temperature gradient along the bar is finally set up, and this gradient then stays constant in space. Typically, such a new steady state gradient is approached exponentially with time after a new temperature-or-heat source or sink, has been introduced. When a "transient conduction" phase is over, heat flow may still continue at high power, so long as temperatures do not change.

An example of transient conduction that does not end with steady-state conduction, but rather no conduction, occurs when a hot copper ball is dropped into oil at a low temperature. Here, the temperature field within the object begins to change as a function of time, as the heat is removed from the metal, and the interest lies in analyzing this spatial change of temperature within the object over time, until all gradients disappear entirely (the ball has reached the same temperature as the oil). Mathematically, this condition is also approached exponentially; in theory it takes infinite time, but in practice it is over, for all intents and purposes, in a much shorter period. At the end of this process with no heat sink but the internal parts of the ball (which are finite), there is no steady state heat conduction to reach. Such a state never occurs in this situation, but rather the end of the process is when there is no heat conduction at all.

The analysis of non steady-state conduction systems is more complex than that of steady-state systems. If the conducting body has a simple shape, then exact analytical mathematical expressions and solutions may be possible. However, most often, because of complicated shapes with varying thermal conductivities within the shape (i.e., most complex objects, mechanisms or machines in engineering) often the application of approximate theories is required, and/or numerical analysis by computer. One popular graphical method involves the use of Heisler Charts.

Occasionally, transient conduction problems may be considerably simplified if regions of the object being heated or cooled can be identified, for which thermal conductivity is very much greater than that for heat paths leading into the region. In this case, the region with high conductivity can often be treated in the lumped capacitance model, as a "lump" of material with a simple thermal capacitance consisting of its aggregate heat capacity. Such regions warm or cool, but show no significant temperature *variation* across their extent, during the process (as compared to the rest of the system). This is due to their far higher conductance. During transient conduction, therefore, the temperature across their conductive regions changes uniformly in space, and as a simple exponential in time. An example of such systems are those that follow Newton's law of cooling during transient cooling (or the reverse during heating). The equivalent thermal circuit consists of a simple capacitor in series with a resistor. In such cases, the remainder of the system with high thermal resistance (comparatively low conductivity) plays the role of the resistor in the circuit.

Relativistic Conduction

The theory of relativistic heat conduction is a model that is compatible with the theory of special relativity. For most of the last century, it was recognized that the Fourier equation is in contradiction with the theory of relativity because it admits an infinite speed of propagation of heat signals. For example, according to the Fourier equation, a pulse of heat at the origin would be felt at infinity instantaneously. The speed of information propagation is faster than the speed of light in vacuum, which is physically inadmissible within the framework of relativity.

Quantum Conduction

Second sound is a quantum mechanical phenomenon in which heat transfer occurs by wave-like motion, rather than by the more usual mechanism of diffusion. Heat takes the place of pressure in normal sound waves. This leads to a very high thermal conductivity. It is known as "second sound" because the wave motion of heat is similar to the propagation of sound in air.

Fourier's Law

The law of heat conduction, also known as Fourier's law, states that the time rate of heat transfer through a material is proportional to the negative gradient in the temperature and to the area, at right angles to that gradient, through which the heat flows. We can state this law in two equivalent forms: the integral form, in which we look at the amount of energy flowing into or out of a body as a whole, and the differential form, in which we look at the flow rates or fluxes of energy locally.

Newton's law of cooling is a discrete analog of Fourier's law, while Ohm's law is the electrical analogue of Fourier's law.

Differential form

The differential form of Fourier's law of thermal conduction shows that the local heat flux density, $\vec{q}$, is equal to the product of thermal conductivity, k, and the negative local temperature gradient, $-\nabla T$. The heat flux density is the amount of energy that flows through a unit area per unit time.

$$\vec{q} = -k\nabla T$$

where (including the SI units)

$\vec{q}$ is the local heat flux density, W·m^{-2}

k is the material's conductivity, W·m^{-1}·K^{-1},

∇T is the temperature gradient, K·m^{-1}.

Integral form

By integrating the differential form over the material's total surface S, we arrive at the integral form of Fourier's law:

$$\frac{\partial Q}{\partial t} = -k \oiint \nabla T \cdot dS$$

Conductance

Writing

$$U = \frac{k}{\Delta x},$$

where U is the conductance, in W/(m² K).

Fourier's law can also be stated as:

$$\frac{\Delta Q}{\Delta t} = UA(-\Delta T).$$

The reciprocal of conductance is resistance, R, given by:

$$.R = \frac{1}{U} = \frac{\Delta x}{k} = \frac{A(-\Delta T)}{\dfrac{\Delta Q}{\Delta t}}.$$

Resistance is additive when several conducting layers lie between the hot and cool regions, because A and Q are the same for all layers. In a multilayer partition, the total conductance is related to the conductance of its layers by:

$$\frac{1}{U} = \frac{1}{U_1} + \frac{1}{U_2} + \frac{1}{U_3} + \cdots$$

So, when dealing with a multilayer partition, the following formula is usually used:

$$\frac{\Delta Q}{\Delta t} = \frac{A(-\Delta T)}{\dfrac{\Delta x_1}{k_1} + \dfrac{\Delta x_2}{k_2} + \dfrac{\Delta x_3}{k_3} + \cdots}.$$

For heat conduction from one fluid to another through a barrier, it is sometimes important to consider the conductance of the thin film of fluid that remains stationary next to the barrier. This thin film of fluid is difficult to quantify because its characteristics depend upon complex conditions of turbulence and viscosity—but when dealing with thin high-conductance barriers it can sometimes be quite significant.

Intensive-property Representation

The previous conductance equations, written in terms of extensive properties, can be reformulated in terms of intensive properties. Ideally, the formulae for conductance should produce a quantity with dimensions independent of distance, like Ohm's Law for electrical resistance, $R = V / I,$, and conductance, $G = I / V..$

From the electrical formula: $R = \rho x / A$, where ρ is resistivity, x is length, and A is cross-sectional area, we have $G = kA / x$, where G is conductance, k is conductivity, x is length, and A is cross-sectional area.

For Heat,

$$U = \frac{kA}{\Delta x},$$

where U is the conductance.

Fourier's law can also be stated as:

$$\dot{Q} = U\,\Delta T$$

The rules for combining resistances and conductances (in series and in parallel) are the same for both heat flow and electric current.

Zeroth Law of Thermodynamics

One statement of the so-called zeroth law of thermodynamics is directly focused on the idea of conduction of heat. Bailyn (1994) writes that "... the zeroth law may be stated:

> All diathermal walls are equivalent."

A diathermal wall is a physical connection between two bodies that allows the passage of heat between them. Bailyn is referring to diathermal walls that exclusively connect two bodies, especially conductive walls.

This statement of the 'zeroth law' belongs to an idealized theoretical discourse, and actual physical walls may have peculiarities that do not conform to its generality.

For example, the material of the wall must not undergo a phase transition, such as evaporation or fusion, at the temperature at which it must conduct heat. But when only thermal equilibrium is considered and time is not urgent, so that the conductivity of the material does not matter too much, one suitable heat conductor is as good as another. Conversely, another aspect of the zeroth law is that, subject again to suitable restrictions, a given diathermal wall is indifferent to the nature of the heat bath to which it is connected. For example, the glass bulb of a thermometer acts as a diathermal wall whether exposed to a gas or to a liquid, provided they do not corrode or melt it.

These indifferences are amongst the defining characteristics of heat transfer. In a sense, they are symmetries of heat transfer.

Thermal Radiation

Thermal radiation is electromagnetic radiation generated by the thermal motion of charged particles in matter. All matter with a temperature greater than absolute zero emits thermal radiation. When the temperature of a body is greater than absolute zero, inter-atomic collisions cause the kinetic energy of the atoms or molecules to change. This results in charge-acceleration and/or dipole oscillation which produces electromagnetic radiation, and the wide spectrum of radiation reflects the wide spectrum of energies and accelerations that occur even at a single temperature.

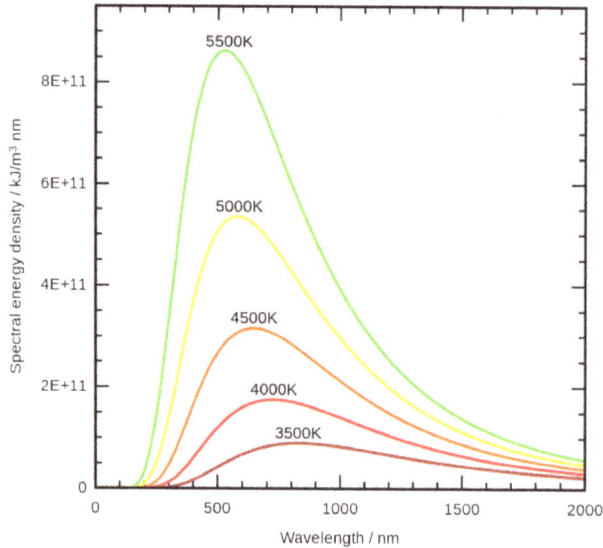

The peak wavelength and total radiated amount vary with temperature according to Wien's displacement law. Although this shows relatively high temperatures, the same relationships hold true for any temperature down to absolute zero. Visible light is between 380 and 750 nm.

Examples of thermal radiation include the visible light and infrared light emitted by an incandescent light bulb, the infrared radiation emitted by animals is detectable with an infrared camera, and the cosmic microwave background radiation. Thermal radiation is different from thermal convection and thermal conduction—a person near a raging bonfire feels radiant heating from the fire, even if the surrounding air is very cold.

Sunlight is part of thermal radiation generated by the hot plasma of the Sun. The Earth also emits thermal radiation, but at a much lower intensity and different spectral distribution (infrared rather than visible) because it is cooler. The Earth's absorption of solar radiation, followed by its outgoing thermal radiation are the two most important processes that determine the temperature and climate of the Earth.

If a radiation-emitting object meets the physical characteristics of a black body in thermodynamic equilibrium, the radiation is called blackbody radiation. Planck's law describes the spectrum of blackbody radiation, which depends only on the object's temperature. Wien's displacement law determines the most likely frequency of the emitted radiation, and the Stefan–Boltzmann law gives the radiant intensity.

Thermal radiation is one of the fundamental mechanisms of heat transfer.

Overview

Thermal radiation is the emission of electromagnetic waves from all matter that has a temperature greater than absolute zero. It represents a conversion of thermal energy into electromagnetic energy. Thermal energy consists of the kinetic energy of random movements of atoms and molecules in matter. All matter with a temperature by definition is composed of particles which have kinetic energy, and which interact with each other. These atoms and molecules are composed of charged particles, i.e., protons and electrons, and kinetic interactions among matter particles result in charge-acceleration and dipole-oscillation. This results in the electrodynamic generation of coupled electric and magnetic fields, resulting in the emission of photons, radiating energy away from the body through its surface boundary. Electromagnetic radiation, including light, does not require the presence of matter to propagate and travels in the vacuum of space infinitely far if unobstructed.

The characteristics of thermal radiation depend on various properties of the surface it is emanating from, including its temperature, its spectral absorptivity and spectral emissive power, as expressed by Kirchhoff's law. The radiation is not monochromatic, i.e., it does not consist of just a single frequency, but comprises a continuous dispersion of photon energies, its characteristic spectrum. If the radiating body and its surface are in thermodynamic equilibrium and the surface has perfect absorptivity at all wavelengths, it is characterized as a black body. A black body is also a perfect emitter. The radiation of such perfect emitters is called black-body radiation. The ratio of any body's emission relative to that of a black body is the body's emissivity, so that a black body has an emissivity of unity.

Absorptivity, reflectivity, and emissivity of all bodies are dependent on the wavelength of the radiation. The temperature determines the wavelength distribution of the electromagnetic radiation. For example, fresh snow, which is highly reflective to visible light (reflectivity about 0.90), appears white due to reflecting sunlight with a peak wavelength of about 0.5 micrometers. Its emissivity at a temperature of about −5 °C (23 °F), peak wavelength of about 12 micrometers, is 0.99.

The distribution of power that a black body emits with varying frequency is described by Planck's law. At any given temperature, there is a frequency f_{max} at which the power emitted is a maximum. Wien's displacement law, and the fact that the frequency is inversely proportional to the wavelength, indicates that the peak frequency f_{max} is proportional to the absolute temperature T of the black body. The photosphere of the sun, at a temperature of approximately 6000 K, emits radiation principally in the (humanly) visible portion of the electromagnetic spectrum. Earth's atmosphere is partly transparent to visible light, and the light reaching the surface is absorbed or reflected. Earth's surface emits the absorbed radiation, approximating the behavior of a black body at 300 K with spectral peak at f_{max}. At these lower frequencies, the atmosphere is largely opaque and radiation from Earth's surface is absorbed or scattered by the atmosphere. Though some radiation escapes into space, most is absorbed and then re-emitted by atmospheric gases. It is this spectral selectivity of the atmosphere that is responsible for the planetary greenhouse effect, contributing to global warming and climate change in general (but also critically contributing to climate stability when the composition and properties of the atmosphere are not changing).

The incandescent light bulb has a spectrum overlapping the black body spectra of the sun and the earth. Some of the photons emitted by a tungsten light bulb filament at 3000 K are in the visible spectrum. Most of the energy is associated with photons of longer wavelengths; these do not help a person see, but still transfer heat to the environment, as can be deduced empirically by observing an incandescent light bulb. Whenever EM radiation is emitted and then absorbed, heat is transferred. This principle is used in microwave ovens, laser cutting, and RF hair removal.

Unlike conductive and convective forms of heat transfer, thermal radiation can be concentrated in a tiny spot by using reflecting mirrors. Concentrating solar power takes advantage of this fact. In many such systems, mirrors are employed to concentrate sunlight into a smaller area. Instead of mirrors, Fresnel lenses can also be used to concentrate heat flux. (In principle, any kind of lens can be used, but only the Fresnel lens design is practical for very large lenses.) Either method can be used to quickly vaporize water into steam using sunlight. For example, the sunlight reflected from mirrors heats the PS10 Solar Power Plant, and during the day it can heat water to 285 °C (558.15 K) or 545 °F.

Surface Effects

Lighter colors and also whites and metallic substances absorb less illuminating light, and thus heat up less; but otherwise color makes small difference as regards heat transfer between an object at everyday temperatures and its surroundings, since the dominant emitted wavelengths are nowhere near the visible spectrum, but rather in the far infrared. Emissivities at those wavelengths have little to do with visual emissivities (visible colors); in the far infra-red, most objects have high emissivities. Thus, except in sunlight, the color of clothing makes little difference as regards warmth; likewise, paint color of houses makes little difference to warmth except when the painted part is sunlit.

The main exception to this is shiny metal surfaces, which have low emissivities both in the visible wavelengths and in the far infrared. Such surfaces can be used to reduce heat transfer in both directions; an example of this is the multi-layer insulation used to insulate spacecraft.

Low-emissivity windows in houses are a more complicated technology, since they must have low emissivity at thermal wavelengths while remaining transparent to visible light.

Nanostructures with spectrally selective thermal emittance properties offer numerous technological applications for energy generation and efficiency, e.g., for cooling photovoltaic cells and buildings. These applications require high emittance in the frequency range corresponding to the atmospheric transparency window in 8 to 13 micron wavelength range. A selective emitter radiating strongly in this range is thus exposed to the clear sky, enabling the use of the outer space as a very low temperature heat sink.

Personalized cooling technology is another example of an application where optical spectral selectivity can be beneficial. Conventional personal cooling is typically achieved through heat conduction and convection. However, the human body is a very efficient emitter of IR radiation, which provides additional cooling mechanism. Most conventional fabrics are opaque to IR radiation and block thermal emission from the body to the environment. Fabrics for personalized cooling applications have been proposed that enable IR transmission to directly pass through clothing, while being opaque at visible wavelengths. Fabrics that are transparent in the infrared can radiate body heat at rates that will significantly reduce the burden on power-hungry air-conditioning systems.

Properties

There are four main properties that characterize thermal radiation (in the limit of the far field):

- Thermal radiation emitted by a body at any temperature consists of a wide range of frequencies. The frequency distribution is given by Planck's law of black-body radiation for an idealized emitter as shown in the diagram at top.

- The dominant frequency (or color) range of the emitted radiation shifts to higher frequencies as the temperature of the emitter increases. For example, a *red hot* object radiates mainly in the long wavelengths (red and orange) of the visible band. If it is heated further, it also begins to emit discernible amounts of green and blue light, and the spread of frequencies in the entire visible range cause it to appear white to the human eye; it is *white hot*. Even at a white-hot temperature of 2000 K, 99% of the energy of the radiation is still in the infrared. This is determined by Wien's displacement law. In the diagram the peak value for each curve moves to the left as the temperature increases.

- The total amount of radiation of all frequencies increases steeply as the temperature rises; it grows as T^4, where T is the absolute temperature of the body. An object at the temperature of a kitchen oven, about twice the room temperature on the absolute temperature scale (600 K vs. 300 K) radiates 16 times as much power per unit area. An object at the temperature of the filament in an incandescent light bulb—roughly 3000 K, or 10 times room temperature—radiates 10,000 times as much energy per unit area. The total radiative intensity of a black body rises as the fourth power of the absolute temperature, as expressed by the Stefan–Boltzmann law. In the plot, the area under each curve grows rapidly as the temperature increases.

- The rate of electromagnetic radiation emitted at a given frequency is proportional to the amount of absorption that it would experience by the source. Thus, a surface that absorbs more red light thermally radiates more red light. This principle applies to all properties of the wave, including wavelength (color), direction, polarization, and even coherence, so that it is quite possible to have thermal radiation which is polarized, coherent, and directional, though polarized and coherent forms are fairly rare in nature far from sources (in terms of wavelength).

Near-field and Far-field

The general properties of thermal radiation as described by the Planck's law apply if the linear dimension of all parts considered, as well as radii of curvature of all surfaces are large compared with the wavelength of the ray considered' (typically from 8-25 micrometres for the emitter at 300 K). Indeed, thermal radiation as discussed above takes only radiating waves (far-field, or electromagnetic radiation) into account. A more sophisticated framework involving electromagnetic theory must be used for smaller distances from the thermal source or surface (near-field thermal radiation). For example, although far-field thermal radiation at distances from surfaces of more than one wavelength is generally not coherent to any extent, near-field thermal radiation (i.e., radiation at distances of a fraction of various radiation wavelengths) may exhibit a degree of both temporal and spatial coherence.

Planck's law of thermal radiation has been challenged in recent decades by predictions and successful demonstrations of the radiative heat transfer between objects separated by nanoscale gaps that deviate significantly from the law predictions. This deviation is especially strong (up to several orders in magnitude) when the emitter and absorber support surface polariton modes that can couple through the gap separating cold and hot objects. However, to take advantage of the surface-polariton-mediated near-field radiative heat transfer, the two objects need to be separated by ultra-narrow gaps on the order of microns or even nanometers. This limitation significantly complicates practical device designs.

Another way to modify the object thermal emission spectrum is by reducing the dimensionality of the emitter itself. This approach builds upon the concept of confining electrons in quantum wells, wires and dots, and tailors thermal emission by engineering confined photon states in two- and three-dimensional potential traps, including wells, wires, and dots. Such spatial confinement concentrates photon states and enhances thermal emission at select frequencies. To achieve the required level of photon confinement, the dimensions of the radiating objects should be on the order of or below the thermal wavelength predicted by Planck's law. Most importantly, the emission spectrum of thermal wells, wires and dots deviates from Planck's law predictions not only in the near field, but also in the far field, which significantly expands the range of their applications.

Subjective Color to the Eye of a Black Body Thermal Radiator

°C (°F)	Subjective color
480 °C (896 °F)	faint red glow
580 °C (1,076 °F)	dark red
730 °C (1,350 °F)	bright red, slightly orange
930 °C (1,710 °F)	bright orange

1,100 °C (2,010 °F)	pale yellowish orange
1,300 °C (2,370 °F)	yellowish white
> 1,400 °C (2,550 °F)	white (yellowish if seen from a distance through atmosphere)

Selected Radiant Heat Fluxes

The time to a damage from exposure to radiative heat is a function of the rate of delivery of the heat. Radiative heat flux and effects: (1 W/cm² = 10 kW/m²)

kW/m²	Effect
170	Maximum flux measured in a post-flashover compartment
80	Thermal Protective Performance test for personal protective equipment
52	Fiberboard ignites at 5 seconds
29	Wood ignites, given time
20	Typical beginning of flashover at floor level of a residential room
16	Human skin: sudden pain and second-degree burn blisters after 5 seconds
12.5	Wood produces ignitable volatiles by pyrolysis
10.4	Human skin: Pain after 3 seconds, second-degree burn blisters after 9 seconds
6.4	Human skin: second-degree burn blisters after 18 seconds
4.5	Human skin: second-degree burn blisters after 30 seconds
2.5	Human skin: burns after prolonged exposure, radiant flux exposure typically encountered during firefighting
1.4	Sunlight, sunburns potentially within 30 minutes

Interchange of Energy

Radiant heat panel for testing precisely quantified energy exposures at National Research Council, near Ottawa, Ontario, Canada

Thermal radiation is one of the principal mechanisms of heat transfer. It entails the emission of a spectrum of electromagnetic radiation due to an object's temperature. Other mechanisms are convection and conduction. The interplay of energy exchange by thermal radiation is characterized by the following equation:

$$\alpha + \rho + \tau = 1.$$

$$\alpha = \epsilon = 1.$$

In a practical situation and room-temperature setting, humans lose considerable energy due to thermal radiation. The energy lost by emitting infrared radiation is partially regained by absorbing the heat flow due to conduction from surrounding objects, and the remainder resulting from generated heat through metabolism. Human skin has an emissivity of very close to 1.0. Using the formulas below shows a human, having roughly 2 square meter in surface area, and a temperature of about 307 K, continuously radiates approximately 1000 watts. If people are indoors, surrounded by surfaces at 296 K, they receive back about 900 watts from the wall, ceiling, and other surroundings, so the net loss is only about 100 watts. These heat transfer estimates are highly dependent on extrinsic variables, such as wearing clothes, i.e. decreasing total thermal circuit conductivity, therefore reducing total output heat flux. Only truly *gray* systems (relative equivalent emissivity/absorptivity and no directional transmissivity dependence in *all* control volume bodies considered) can achieve reasonable steady-state heat flux estimates through the Stefan-Boltzmann law. Encountering this "ideally calculable" situation is almost impossible (although common engineering procedures surrender the dependency of these unknown variables and "assume" this to be the case). Optimistically, these "gray" approximations will get close to real solutions, as most divergence from Stefan-Boltzmann solutions is very small (especially in most STP lab controlled environments).

If objects appear white (reflective in the visual spectrum), they are not necessarily equally reflective (and thus non-emissive) in the thermal infrared. Most household radiators are painted white but this is sensible given that they are not hot enough to radiate any significant amount of heat, and are not designed as thermal radiators at all - they are actually convectors, and painting them matt black would make little difference to their efficacy. Acrylic and urethane based white paints have 93% blackbody radiation efficiency at room temperature (meaning the term "black body" does not always correspond to the visually perceived color of an object). These materials that do not follow the "black color = high emissivity/absorptivity" caveat will most likely have functional spectral emissivity/absorptivity dependence.

Calculation of radiative heat transfer between groups of object, including a 'cavity' or 'surroundings' requires solution of a set of simultaneous equations using the radiosity method. In these calculations, the geometrical configuration of the problem is distilled to a set of numbers called view factors, which give the proportion of radiation leaving any given surface that hits another specific surface. These calculations are important in the fields of solar thermal energy, boiler and furnace design and raytraced computer graphics.

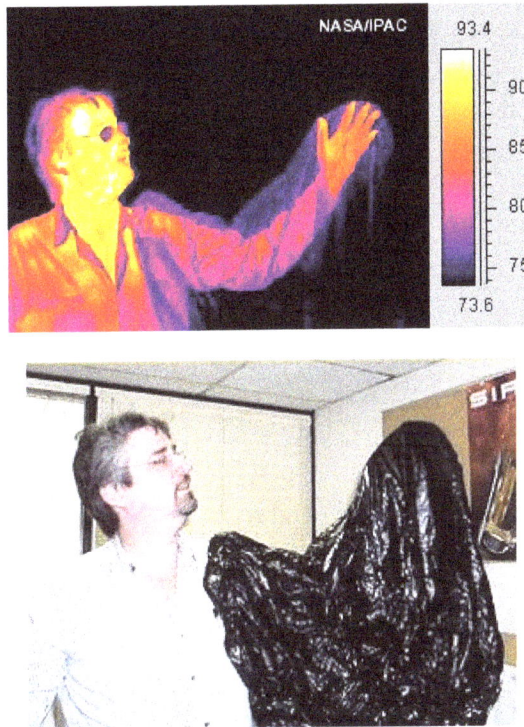

A comparison of a thermal image (top) and an ordinary photograph (bottom) shows that a trash bag is transparent but glass (the man's spectacles) is opaque in long-wavelength infrared.

A selective surface can be used when energy is being extracted from the sun. For instance, when a green house is made, most of the roof and walls are made out of glass. Glass is transparent in the visible (approximately $0.4\ \mu m < \lambda < 0.8\ \mu m$) and near-infrared wavelengths, but opaque to mid- to far-wavelength infrared (approximately $\lambda > 3\ \mu m$). Therefore, glass lets in radiation in the visible range, allowing us to be able to see through it, but does not let out radiation that is emitted from objects at or close to room temperature. This traps what we feel as heat. This is known as the greenhouse effect and can be observed by getting into a car that has been sitting in the sun. Selective surfaces can also be used on solar collectors. We can find out how much help a selective surface coating is by looking at the equilibrium temperature of a plate that is being heated through solar radiation. If the plate is receiving a solar irradiation of 1350 W/m² (minimum is 1325 W/m² on July 4 and maximum is 1418 W/m² on January 3) from the sun the temperature of the plate where the radiation leaving is equal to the radiation being received by the plate is 393 K (248 °F). If the plate has a selective surface with an emissivity of 0.9 and a cut off wavelength of 2.0 μm, the equilibrium temperature is approximately 1250 K (1790 °F). The calculations were made neglecting convective heat transfer and neglecting the solar irradiation absorbed in the clouds/atmosphere for simplicity, the theory is still the same for an actual problem.

To reduce the heat transfer from a surface, such as a glass window, a clear reflective film with a low emissivity coating can be placed on the interior of the surface. "Low-emittance (low-E) coatings are microscopically thin, virtually invisible, metal or metallic oxide layers deposited on a window or skylight glazing surface primarily to reduce the U-factor by suppressing radiative heat flow". By adding this coating we are limiting the amount of radiation that leaves the window thus increasing the amount of heat that is retained inside the window.

Radiative Power

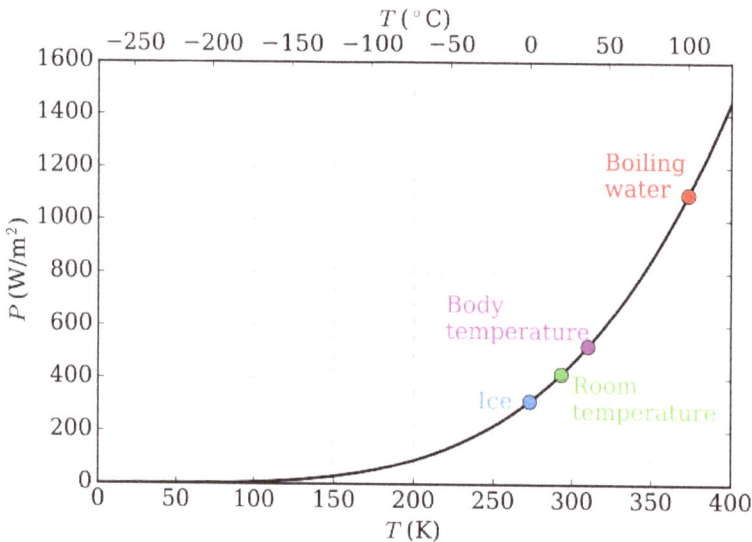

Power emitted by a black body plotted against the temperature according to the Stefan–Boltzmann law.

Thermal radiation power of a black body per unit area of radiating surface per unit of solid angle and per unit frequency v is given by Planck's law as:

$$u(v,T) = \frac{2hv^3}{c^2} \cdot \frac{1}{e^{hv/k_B T} - 1}$$

This formula mathematically follows from calculation of spectral distribution of energy in quantized electromagnetic field which is in complete thermal equilibrium with the radiating object. The equation is derived as an infinite sum over all possible frequencies. The energy, , of each photon is multiplied by the number of states available at that frequency, and the probability that each of those states will be occupied.

Thermal Grease

Thermal grease (also called CPU grease, heat paste, heat sink compound, heat sink paste, thermal compound, thermal gel, thermal interface material, or thermal paste) is a kind of thermally conductive (but usually electrically insulating) compound, which is commonly used as an interface between heat sinks and heat sources (e.g., high-power semiconductor devices). The main role of thermal grease is to eliminate air gaps or spaces (which act as thermal insulator) from the interface area so as to maximize heat transfer. Thermal grease is an example of a Thermal interface material.

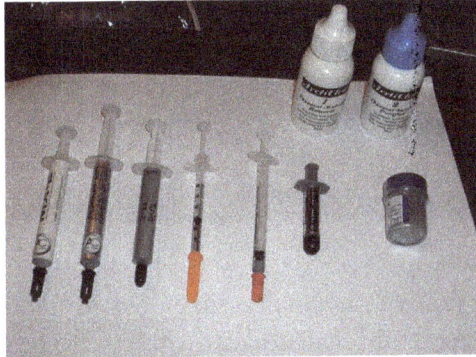

From left to right: Arctic Cooling MX-2 and MX-3, Tuniq TX-3, Cool Laboratory Liquid Metal Pro(Liquid Metal based), Shin-Etsu MicroSi G751, Arctic Silver 5, Powdered Diamond. In background Arctic Silver grease remover

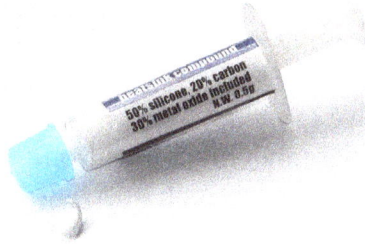

Silicone thermal compound

Metal (silver) thermal compound

Metal thermal grease applied to a chip

Surface imperfections

As opposed to thermal adhesive, thermal grease does not add mechanical strength to the bond between heat source and heat sink. It will have to be coupled with a mechanical fixation mechanism such as screws, allowing for pressure between the two, spreading the thermal grease onto the heat source.

Composition

Thermal grease consists of a polymerizable liquid matrix and large volume fractions of electrically insulating, but thermally conductive filler. Typical matrix materials are epoxies, silicones, urethanes, and acrylates, solvent-based systems, hot-melt adhesives, and pressure-sensitive adhesive tapes are also available. Aluminum oxide, boron nitride, zinc oxide, and increasingly aluminum nitride are used as fillers for these types of adhesives. The filler loading can be as high as 70–80 wt %, and the fillers raise the thermal conductivity of the base matrix from 0.17–0.3 watts per meter Kelvin or W/(m·K), up to about 2 W/(m·K).

Silver thermal compounds may have a conductivity of 3 to 8 W/(m·K) or more. However, metal-based thermal grease can be electrically conductive and capacitive; if some flows onto the circuits it can cause malfunctioning and damage.

Filler Properties

Compound	Thermal conductivity (ca. 300 K) (W m^{-1} K^{-1})	Electrical resistivity (ca. 300 K) (Ω cm)	Thermal expansion coefficient (10^{-6} K^{-1})	Reference
Diamond	20 – 2000	$10^{16} - 10^{20}$	0.8 (15 – 150 °C)	
Silver	418		1.465 (0 °C)	
Aluminum nitride	100 – 170	$> 10^{11}$	3.5 (300 – 600 K)	
β-Boron nitride	100	$> 10^{10}$	4.9	
Zinc oxide	25.2			

Thermal Mass

Effect of Heavy-weight and Light-weight constructions on the internal temperature of a naturally ventilated school classroom

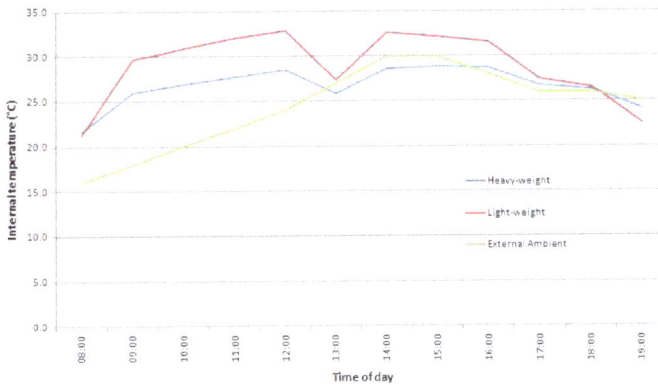

The benefit of thermal mass is shown in this comparison of how heavy and light weight constructions affect internal temperature

In building design, thermal mass is a property of the mass of a building which enables it to store heat, providing "inertia" against temperature fluctuations. It is sometimes known as the thermal flywheel effect. For example, when outside temperatures are fluctuating throughout the day, a large thermal mass within the insulated portion of a house can serve to "flatten out" the daily temperature fluctuations, since the thermal mass will absorb thermal energy when the surroundings are higher in temperature than the mass, and give thermal energy back when the surroundings are cooler, without reaching thermal equilibrium. This is distinct from a material's insulative value, which reduces a building's thermal conductivity, allowing it to be heated or cooled relatively separate from the outside, or even just retain the occupants' thermal energy longer.

Scientifically, thermal mass is equivalent to thermal capacitance or heat capacity, the ability of a body to store thermal energy. It is typically referred to by the symbol C_{th} and measured in units of J/°C or J/K (which are equivalent). Thermal mass may also be used for bodies of water, machines or machine parts, living things, or any other structure or body in engineering or biology. In those contexts, the term "heat capacity" is typically used instead.

Background

The equation relating thermal energy to thermal mass is:

$$Q = C_{th}\Delta T$$

where Q is the thermal energy transferred, C_{th} is the thermal mass of the body, and ΔT is the change in temperature.

For example, if 250 J of heat energy is added to a copper gear with a thermal mass of 38.46 J/°C, its temperature will rise by 6.50 °C. If the body consists of a homogeneous material with sufficiently known physical properties, the thermal mass is simply the mass of material present times the specific heat capacity of that material. For bodies made of many materials, the sum of heat capacities for their pure components may be used in the calculation, or in some cases (as for a whole animal, for example) the number may simply be measured for the entire body in question, directly.

As an extensive property, heat capacity is characteristic of an object; its corresponding intensive property is specific heat capacity, expressed in terms of a measure of the amount of material such as mass or number of moles, which must be multiplied by similar units to give the heat capacity of the entire body of material. Thus the heat capacity can be equivalently calculated as the product of the mass m of the body and the specific heat capacity c for the material, or the product of the number of moles of molecules present n and the molar specific heat capacity . For discussion of *why* the thermal energy storage abilities of pure substances vary.

For a body of uniform composition, C_{th} can be approximated by

$$C_{th} = mc_p$$

where m is the mass of the body and c_p is the isobaric specific heat capacity of the material averaged over temperature range in question. For bodies composed of numerous different materials, the thermal masses for the different components can just be added together.

Thermal Mass in Buildings

Thermal mass is effective in improving building comfort in any place that experiences these types of daily temperature fluctuations—both in winter as well as in summer. When used well and combined with passive solar design, thermal mass can play an important role in major reductions to energy use in active heating and cooling systems. The terms *heavy-weight* and *light-weight* are often used to describe buildings with different thermal mass strategies, and affects the choice of numerical factors used in subsequent calculations to describe their thermal response to heating and cooling. In building services engineering, the use of dynamic simulation computational modelling software has allowed for the accurate calculation of the environmental performance within buildings with different constructions and for different annual climate data sets. This allows the architect or engineer to explore in detail the relationship between heavy-weight and light-weight constructions, as well as insulation levels, in reducing energy consumption for mechanical heating or cooling systems, or even removing the need for such systems altogether.

Properties Required for Good Thermal Mass

Ideal materials for thermal mass are those materials that have:

- high specific heat capacity,

- high density

Any solid, liquid, or gas that has mass will have some thermal mass. A common misconception is that only concrete or earth soil has thermal mass; even air has thermal mass (although very little).

A table of volumetric heat capacity for building materials is available here, but note that their definition of thermal mass is slightly different.

Use of Thermal Mass in Different Climates

The correct use and application of thermal mass is dependent on the prevailing climate in a district.

Temperate and Cold Temperate Climates

Solar-exposed thermal mass

Thermal mass is ideally placed within the building and situated where it still can be exposed to low-angle winter sunlight (via windows) but insulated from heat loss. In summer the same thermal mass should be obscured from higher-angle summer sunlight in order to prevent overheating of the structure.

The thermal mass is warmed passively by the sun or additionally by internal heating systems during the day. Thermal energy stored in the mass is then released back into the interior during the night. It is essential that it be used in conjunction with the standard principles of passive solar design.

Any form of thermal mass can be used. A concrete slab foundation either left exposed or covered with conductive materials, e.g. tiles, is one easy solution. Another novel method is to place the masonry facade of a timber-framed house on the inside ('reverse-brick veneer'). Thermal mass in this situation is best applied over a large area rather than in large volumes or thicknesses. 7.5–10 cm (3-4") is often adequate.

Since the most important source of thermal energy is the Sun, the ratio of glazing to thermal mass is an important factor to consider. Various formulas have been devised to determine this. As a general rule, additional solar-exposed thermal mass needs to be applied in a ratio from 6:1 to 8:1 for any area of sun-facing (north-facing in Southern Hemisphere or south-facing in Northern Hemisphere) glazing above 7% of the total floor area. For example, a 200 m² house with 20 m² of sun-facing glazing has 10% of glazing by total floor area; 6 m² of that glazing will require additional thermal mass. Therefore, using the 6:1 to 8:1 ratio above, an additional 36–48 m² of solar-exposed thermal mass is required. The exact requirements vary from climate to climate.

A modern school classroom with natural ventilation by opening windows and exposed thermal mass from a solid concrete floor soffit to help control summertime temperatures

Thermal Mass for Limiting Summertime Overheating

Thermal mass is ideally placed within a building where it is shielded from direct solar gain but exposed to the building occupants. It is therefore most commonly associated with solid concrete floor slabs in naturally ventilated or low-energy mechanically ventilated buildings where the concrete soffit is left exposed to the occupied space.

During the day heat is gained from the sun, the occupants of the building, and any electrical lighting and equipment, causing the air temperatures within the space to increase, but this heat is absorbed by the exposed concrete slab above, thus limiting the temperature rise within the space to be within acceptable levels for human thermal comfort. In addition the lower surface temperature of the concrete slab also absorbs radiant heat directly from the occupants, also benefiting their thermal comfort.

By the end of the day the slab has in turn warmed up, and now, as external temperatures decrease, the heat can be released and the slab cooled down, ready for the start of the next day. However this "regeneration" process is only effective if the building ventilation system is operated at night to carry away the heat from the slab. In naturally ventilated buildings it is normal to provide automated window openings to facilitate this process automatically.

Hot, Arid Climates (e.g. Desert)

This is a classical use of thermal mass. Examples include adobe or rammed earth houses. Its function is highly dependent on marked diurnal temperature variations. The wall predominantly acts to retard heat transfer from the exterior to the interior during the day. The high volumetric heat capacity and thickness prevents thermal energy from reaching the inner surface. When temperatures fall at night, the walls re-radiate the thermal energy back into the night sky. In this application it is important for such walls to be massive to prevent heat transfer into the interior.

An adobe walled building in Santa Fe, New Mexico

Hot Humid Climates (e.g. Sub-tropical and Tropical)

The use of thermal mass is the most challenging in this environment where night temperatures remain elevated. Its use is primarily as a temporary heat sink. However, it needs to be strategically located to prevent overheating. It should be placed in an area that is not directly exposed to solar gain and also allows adequate ventilation at night to carry away stored energy without increasing internal temperatures any further. If to be used at all it should be used in judicious amounts and again not in large thicknesses.

Cold incoming tap water may be piped through radiators to draw summer thermal energy from the air. In most areas, its initial temperature is 60 °F (16 °C) degrees. Since the existing plumbing is deep underground, it's well insulated from the heat of the day.

Materials Commonly used for Thermal Mass

- Water: water has the highest volumetric heat capacity of all commonly used material. Typically, it is placed in large container(s), acrylic tubes for example, in an area with direct sunlight. It may also be used to saturate other types material such as soil to increase heat capacity.

- Concrete, clay bricks and other forms of masonry: the thermal conductivity of concrete depends on its composition and curing technique. Concretes with stones are more thermally conductive than concretes with ash, perlite, fibers, and other insulating aggregates.

- Insulated concrete panels consist of an inner layer of concrete to provide the thermal mass factor. This is insulated from the outside by a conventional foam insulation and then covered again with an outer layer of concrete. The effect is a highly efficient building insulation envelope.

- Insulating concrete forms are commonly used to provide thermal mass to building structures. Insulating concrete forms provide the specific heat capacity and mass of concrete. Thermal inertia of the structure is very high because the mass is insulated on both sides.

- Clay brick, adobe brick or mudbrick.

- Earth, mud and sod: dirt's heat capacity depends on its density, moisture content, particle shape, temperature, and composition. Early settlers to Nebraska built houses with thick walls made of dirt and sod because wood, stone, and other building materials were scarce. The extreme thickness of the walls provided some insulation, but mainly served as thermal mass, absorbing thermal energy during the day and releasing it during the night. Nowadays, people sometimes use earth sheltering around their homes for the same effect. In earth sheltering, the thermal mass comes not only from the walls of the building, but from the surrounding earth that is in physical contact with the building. This provides a fairly constant, moderating temperature that reduces heat flow through the adjacent wall.

- Rammed earth: rammed earth provides excellent thermal mass because of its high density, and the high specific heat capacity of the soil used in its construction.

- Natural rock and stone.

- Logs are used as a building material to create the exterior, and perhaps also the interior, walls of homes. Log homes differ from some other construction materials listed above because solid wood has both moderate R-value (insulation) and also significant thermal mass. In contrast, water, earth, rocks, and concrete all have low R-values.

- Phase-change materials

Seasonal Energy Storage

If enough mass is used it can create a seasonal advantage. That is, it can heat in the winter and cool in the summer. This is sometimes called passive annual heat storage or PAHS. The PAHS system has been successfully used at 7000 ft. in Colorado and in a number of homes in Montana.The Earthships of New Mexico utilize passive heating and cooling as well as using recycled tires for foundation wall yielding a maximum PAHS/STES. It has also been used successfully in the UK at Hockerton Housing Project.

Thermal Expansion

Thermal expansion is the tendency of matter to change in shape, area, and volume in response to a change in temperature, through heat transfer.

Temperature is a monotonic function of the average molecular kinetic energy of a substance. When a substance is heated, the kinetic energy of its molecules increases. Thus, the molecules begin moving more and usually maintain a greater average separation. Materials which contract with increasing temperature are unusual; this effect is limited in size, and only occurs within limited temperature ranges. The degree of expansion divided by the change in temperature is called the material's coefficient of thermal expansion and generally varies with temperature.

Expansion joint in a road bridge used to avoid damage from thermal expansion.

Overview

Predicting Expansion

If an equation of state is available, it can be used to predict the values of the thermal expansion at all the required temperatures and pressures, along with many other state functions.

Contraction Effects (Negative Thermal Expansion)

A number of materials contract on heating within certain temperature ranges; this is usually called negative thermal expansion, rather than "thermal contraction". For example, the coefficient of thermal expansion of water drops to zero as it is cooled to 3.983 °C and then becomes negative below this temperature; this means that water has a maximum density at this temperature, and this leads to bodies of water maintaining this temperature at their lower depths during extended periods of sub-zero weather. Also, fairly pure silicon has a negative coefficient of thermal expansion for temperatures between about 18 and 120 Kelvin.

Factors Affecting Thermal Expansion

Unlike gases or liquids, solid materials tend to keep their shape when undergoing thermal expansion.

Thermal expansion generally decreases with increasing bond energy, which also has an effect on the melting point of solids, so, high melting point materials are more likely to have lower thermal expansion. In general, liquids expand slightly more than solids. The thermal expansion of glasses is higher compared to that of crystals. At the glass transition temperature, rearrangements that occur in an amorphous material lead to characteristic discontinuities of coefficient of thermal expansion or specific heat. These discontinuities allow detection of the glass transition temperature where a supercooled liquid transforms to a glass.

Absorption or desorption of water (or other solvents) can change the size of many common materials; many organic materials change size much more due to this effect than they do to thermal expansion. Common plastics exposed to water can, in the long term, expand by many percent.

Coefficient of Thermal Expansion

The coefficient of thermal expansion describes how the size of an object changes with a change in temperature. Specifically, it measures the fractional change in size per degree change in temperature at a constant pressure. Several types of coefficients have been developed: volumetric, area, and linear. Which is used depends on the particular application and which dimensions are considered important. For solids, one might only be concerned with the change along a length, or over some area.

The volumetric thermal expansion coefficient is the most basic thermal expansion coefficient, and the most relevant for fluids. In general, substances expand or contract when their temperature changes, with expansion or contraction occurring in all directions. Substances that expand at the same rate in every direction are called isotropic. For isotropic materials, the area and volumetric thermal expansion coefficient are, respectively, approximately twice and three times larger than the linear thermal expansion coefficient.

Mathematical definitions of these coefficients are defined below for solids, liquids, and gases.

General Volumetric Thermal Expansion Coefficient

In the general case of a gas, liquid, or solid, the volumetric coefficient of thermal expansion is given by

$$\alpha_V = \frac{1}{V} \left(\frac{\partial V}{\partial T} \right)_p$$

The subscript p indicates that the pressure is held constant during the expansion, and the subscript V stresses that it is the volumetric (not linear) expansion that enters this general definition. In the case of a gas, the fact that the pressure is held constant is important, because the volume of a gas will vary appreciably with pressure as well as temperature. For a gas of low density this can be seen from the ideal gas law.

Expansion in Solids

When calculating thermal expansion it is necessary to consider whether the body is free to expand or is constrained. If the body is free to expand, the expansion or strain resulting from an increase in temperature can be simply calculated by using the applicable coefficient of thermal expansion.

If the body is constrained so that it cannot expand, then internal stress will be caused (or changed) by a change in temperature. This stress can be calculated by considering the strain that would occur if the body were free to expand and the stress required to reduce that strain to zero, through the stress/strain relationship characterised by the elastic or Young's modulus. In the special case of solid materials, external ambient pressure does not usually appreciably affect the size of an object and so it is not usually necessary to consider the effect of pressure changes.

Common engineering solids usually have coefficients of thermal expansion that do not vary significantly over the range of temperatures where they are designed to be used, so where extremely high accuracy is not required, practical calculations can be based on a constant, average, value of the coefficient of expansion.

Linear Expansion

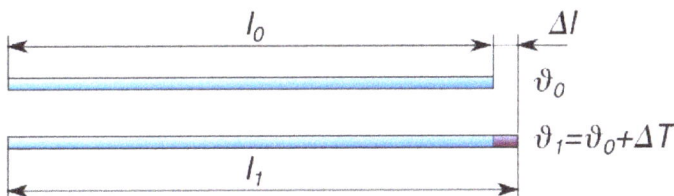

Change in length of a rod due to thermal expansion.

Linear expansion means change in one dimension (length) as opposed to change in volume (volumetric expansion). To a first approximation, the change in length measurements of an object due to thermal expansion is related to temperature change by a "linear expansion coefficient". It is the fractional change in length per degree of temperature change. Assuming negligible effect of pressure, we may write:

$$\alpha_L = \frac{1}{L}\frac{dL}{dT}$$

where L is a particular length measurement and dL/dT is the rate of change of that linear dimension per unit change in temperature.

The change in the linear dimension can be estimated to be:

$$\frac{\Delta L}{L} = \alpha_L \Delta T$$

Effects on Strain

For solid materials with a significant length, like rods or cables, an estimate of the amount of thermal expansion can be described by the material strain, given by $\epsilon_{\text{thermal}}$ and defined as:

$$\epsilon_{\text{thermal}} = \frac{(L_{\text{final}} - L_{\text{initial}})}{L_{\text{initial}}}$$

where L_{initial} is the length before the change of temperature and L_{final} is the length after the change of temperature.

Expansion in Mixtures and Alloys

The expansivity of the components of the mixture can cancel each other like in invar.

The thermal expansivity of a mixture from the expansivities of the pure components and their excess expansivities follow from:

$$\frac{\partial V}{\partial T} = \sum_i \frac{\partial V_i}{\partial T} + \sum_i \frac{\partial V_i^E}{\partial T}$$

$$\alpha = \sum_i \alpha_i V_i + \sum_i \alpha_i^E V_i^E$$

$$\frac{\partial \overline{V}^E{}_i}{\partial T} = R\frac{\partial(ln(\gamma_i))}{\partial P} + RT\frac{\partial^2}{\partial T \partial P}ln(\gamma_i)$$

Apparent and Absolute Expansion

When measuring the expansion of a liquid, the measurement must account for the expansion of the container as well. For example, a flask that has been constructed with a long narrow stem filled with enough liquid that the stem itself is partially filled, when placed in a heat bath will initially show the column of liquid in the stem to drop followed by the immediate increase of that column until the flask-liquid-heat bath system has thermalized. The initial observation of the column of liquid dropping is not due to an initial contraction of the liquid but rather the expansion of the flask as it contacts the heat bath first. Soon after, the liquid in the flask is heated by the flask itself and begins to expand. Since liquids typically have a greater expansion over solids, the liquid in the flask eventually exceeds that of the flask, causing the column of liquid in the flask to rise. A direct measurement of the height of the liquid column is a measurement of the apparent expansion of the liquid. The absolute expansion of the liquid is the apparent expansion corrected for the expansion of the containing vessel.

Examples and Applications

Thermal expansion of long continuous sections of rail tracks is the driving force for rail buckling. This phenomenon resulted in 190 train derailments during 1998–2002 in the US alone.

The expansion and contraction of materials must be considered when designing large structures, when using tape or chain to measure distances for land surveys, when designing molds for casting hot material, and in other engineering applications when large changes in dimension due to temperature are expected.

Thermal expansion is also used in mechanical applications to fit parts over one another, e.g. a bushing can be fitted over a shaft by making its inner diameter slightly smaller than the diameter of the shaft, then heating it until it fits over the shaft, and allowing it to cool after it has been pushed over the shaft, thus achieving a 'shrink fit'. Induction shrink fitting is a common industrial method to pre-heat metal components between 150 °C and 300 °C thereby causing them to expand and allow for the insertion or removal of another component.

There exist some alloys with a very small linear expansion coefficient, used in applications that demand very small changes in physical dimension over a range of temperatures. One of these is Invar 36, with α approximately equal to 0.6×10^{-6} K^{-1}. These alloys are useful in aerospace applications where wide temperature swings may occur.

Pullinger's apparatus is used to determine the linear expansion of a metallic rod in the laboratory. The apparatus consists of a metal cylinder closed at both ends (called a steam jacket). It is provided with an inlet and outlet for the steam. The steam for heating the rod is supplied by a boiler which is connected by a rubber tube to the inlet. The center of the cylinder contains a hole to insert a thermometer. The rod under investigation is enclosed in a steam jacket. One of its ends is free, but the other end is pressed against a fixed screw. The position of the rod is determined by a micrometer screw gauge or spherometer.

Drinking glass with fracture due to uneven thermal expansion after pouring of hot liquid into the otherwise cool glass

The control of thermal expansion in brittle materials is a key concern for a wide range of reasons. For example, both glass and ceramics are brittle and uneven temperature causes uneven expansion which again causes thermal stress and this might lead to fracture. Ceramics need to be joined or work in consort with a wide range of materials and therefore their expansion must be matched to the application. Because glazes need to be firmly attached to the underlying porcelain (or other body type) their thermal expansion must be tuned to 'fit' the body so that crazing or shivering do not occur. Good example of products whose thermal expansion is the key to their success are Corning-Ware and the spark plug. The thermal expansion of ceramic bodies can be controlled by firing to create crystalline species that will influence the overall expansion of the material in the desired direction. In addition or instead the formulation of the body can employ materials delivering particles of the desired expansion to the matrix. The thermal expansion of glazes is controlled by their chemical composition and the firing schedule to which they were subjected. In most cases there are complex issues involved in controlling body and glaze expansion, adjusting for thermal expansion must be done with an eye to other properties that will be affected, generally trade-offs are required.

Thermal expansion can have a noticeable effect in gasoline stored in above ground storage tanks which can cause gasoline pumps to dispense gasoline which may be more compressed than gasoline held in underground storage tanks in the winter time or less compressed than gasoline held in underground storage tanks in the summer time.

Heat-induced expansion has to be taken into account in most areas of engineering. A few examples are:

- Metal framed windows need rubber spacers

- Rubber tires

- Metal hot water heating pipes should not be used in long straight lengths

- Large structures such as railways and bridges need expansion joints in the structures to avoid sun kink

- One of the reasons for the poor performance of cold car engines is that parts have inefficiently large spacings until the normal operating temperature is achieved.

- A gridiron pendulum uses an arrangement of different metals to maintain a more temperature stable pendulum length.

- A power line on a hot day is droopy, but on a cold day it is tight. This is because the metals expand under heat.

- Expansion joints that absorb the thermal expansion in a piping system.

- Precision engineering nearly always requires the engineer to pay attention to the thermal expansion of the product. For example, when using a scanning electron microscope even small changes in temperature such as 1 degree can cause a sample to change its position relative to the focus point.

Thermometers are another application of thermal expansion – most contain a liquid (usually mercury or alcohol) which is constrained to flow in only one direction (along the tube) due to changes in volume brought about by changes in temperature. A bi-metal mechanical thermometer uses a bimetallic strip and bends due to the differing thermal expansion of the two metals.

Metal pipes made of different materials are heated by passing steam through them. While each pipe is being tested, one end is securely fixed and the other rests on a rotating shaft, the motion of which is indicated with a pointer. The linear expansion of the different metals is compared qualitatively and the coefficient of linear thermal expansion is calculated.

Thermal Expansion Coefficients for Various Materials

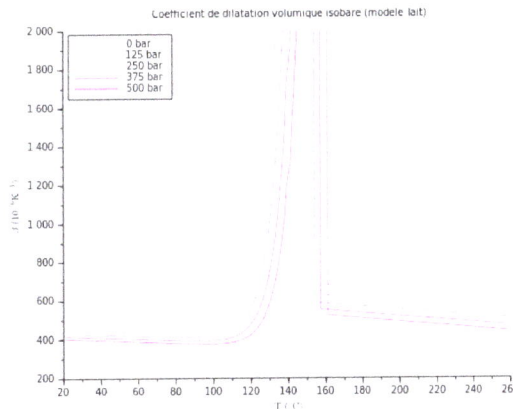

Volumetric thermal expansion coefficient for a semicrystalline polypropylene.

This section summarizes the coefficients for some common materials.

For isotropic materials the coefficients linear thermal expansion α and volumetric thermal expansion α_V are related by $\alpha_V = 3\alpha$. For liquids usually the coefficient of volumetric expansion is listed and linear expansion is calculated here for comparison.

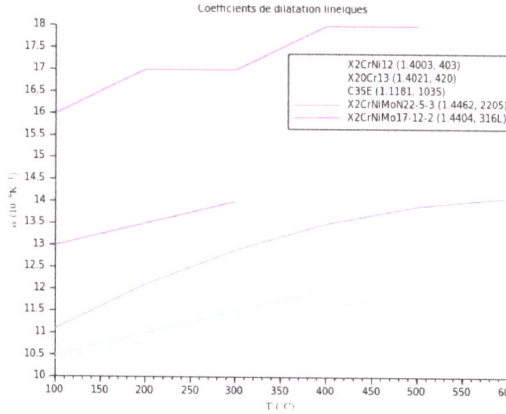

Linear thermal expansion coefficient for some steel grades.

For common materials like many metals and compounds, the thermal expansion coefficient is inversely proportional to the melting point. In particular for metals the relation is:

$$\alpha \approx \frac{0.020}{M_P}$$

for halides and oxides

$$\alpha \approx \frac{0.038}{M_P} - 7.0 \cdot 10^{-6} \, \text{K}^{-1}$$

In the table below, the range for α is from 10^{-7} K^{-1} for hard solids to 10^{-3} K^{-1} for organic liquids. The coefficient α varies with the temperature and some materials have a very high variation ; see for example the variation vs. temperature of the volumetric coefficient for a semicrystalline polypropylene (PP) at different pressure, and the variation of the linear coefficient vs. temperature for some steel grades (from bottom to top: ferritic stainless steel, martensitic stainless steel, carbon steel, duplex stainless steel, austenitic steel).

(The formula $\alpha_v \approx 3\alpha$ is usually used for solids.)

Material	Linear coefficient α at 20 °C (10^{-6} K^{-1})	Volumetric coefficient α_v at 20 °C (10^{-6} K^{-1})	Notes
Aluminium	23.1	69	
Aluminium nitride	5.3	4.2	
Benzocyclobutene	42	126	
Brass	19	57	
Carbon steel	10.8	32.4	

CFRP	− 0.8	Anisotropic	Fiber direction
Concrete	12	36	
Copper	17	51	
Diamond	1	3	
Ethanol	250	750	
Gallium(III) arsenide	5.8	17.4	
Gasoline	317	950	
Glass	8.5	25.5	
Glass, borosilicate	3.3	9.9	matched sealing partner for tungsten, molybdenum and kovar.
Glass (Pyrex)	3.2		
Glycerine		485	
Gold	14	42	
Helium		36.65	
Indium phosphide	4.6	13.8	
Invar	1.2	3.6	
Iron	11.8	33.3	
Kapton	20	60	DuPont Kapton 200EN
Lead	29	87	
Macor	9.3		
Magnesium	26	78	
Mercury	61	182	
Molybdenum	4.8	14.4	
Nickel	13	39	
Oak	54		Perpendicular to the grain
Douglas-fir	27	75	radial
Douglas-fir	45	75	tangential
Douglas-fir	3.5	75	parallel to grain
Platinum	9	27	
PP	150	450	
PVC	52	156	
Quartz (fused)	0.59	1.77	
Quartz	0.33	1	
Rubber	*disputed*	*disputed*	
Sapphire	5.3		Parallel to C axis, or [001]
Silicon Carbide	2.77	8.31	
Silicon	2.56	9	
Silver	18	54	
Sitall	0±0.15	0±0.45	average for −60 °C to 60 °C
Stainless steel	10.1 ~ 17.3	30.3 ~ 51.9	

Steel	11.0 ~ 13.0	33.0 ~ 39.0	Depends on composition
Titanium	8.6	26	
Tungsten	4.5	13.5	
Turpentine		90	
Water	69	207	
YbGaGe	$\doteq 0$	$\doteq 0$	Refuted
Zerodur	≈ 0.02		at 0...50 °C

Heat

In physics, heat is energy that spontaneously passes between a system and its surroundings in some way other than through work or the transfer of matter. When a suitable physical pathway exists, heat flows spontaneously from a hotter to a colder body. The transfer can be by contact between the source and the destination body, as in conduction; or by radiation between remote bodies; or by conduction and radiation through a thick solid wall; or by way of an intermediate fluid body, as in convective circulation; or by a combination of these.

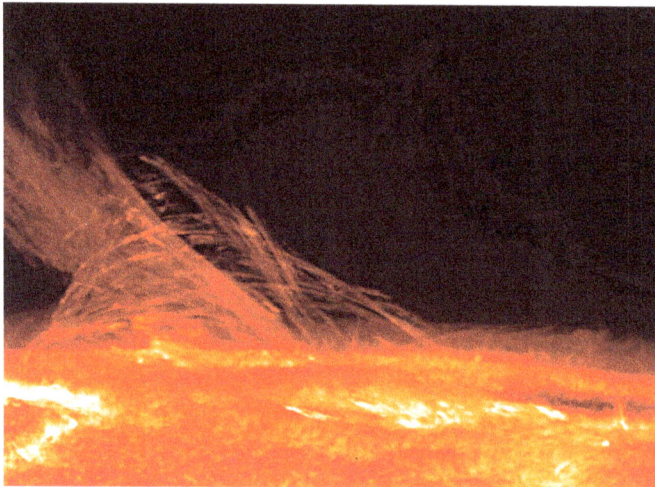

The Sun and Earth form an ongoing example of a heating process. Some of the Sun's thermal radiation strikes and heats the Earth. Compared to the Sun, Earth has a much lower temperature and so sends far less thermal radiation back to the Sun. The heat of this process can be quantified by the net amount, and direction (Sun to Earth), of energy it transferred in a given period of time.

Because heat refers to a quantity of energy transferred between two bodies, it is not a state function of either of the bodies, in contrast to temperature and internal energy. Instead, according to the first law of thermodynamics heat exchanged during some process contributes to the change in the internal energy, and the amount of heat can be quantified by the equivalent amount of work that would bring about the same change.

While heat flows spontaneously from hot to cold, it is possible to construct a heat pump or refrigeration system that does work to increase the difference in temperature between two systems. Conversely, a heat engine reduces an existing temperature difference to do work on another system.

Historically, many energy units for measurement of heat have been used. The standards-based unit in the International System of Units (SI) is the joule (J). Heat is measured by its effect on the states of interacting bodies, for example, by the amount of ice melted or a change in temperature. The quantification of heat via the temperature change of a body is called calorimetry, and is widely used in practice. In calorimetry, sensible heat is defined with respect to a specific chosen state variable of the system, such as pressure or volume. Sensible heat causes a change of the temperature of the system while leaving the chosen state variable unchanged. Heat transfer that occurs at a constant system temperature but changes the state variable is called latent heat with respect to the variable. For infinitesimal changes, the total incremental heat transfer is then the sum of the latent and sensible heat.

History

Physicist James Clerk Maxwell, in his 1871 classic *Theory of Heat*, was one of many who began to build on the already established idea that heat has something to do with matter in motion. This was the same idea put forth by Benjamin Thompson in 1798, who said he was only following up on the work of many others. One of Maxwell's recommended books was *Heat as a Mode of Motion*, by John Tyndall. Maxwell outlined four stipulations for the definition of heat:

- It is *something which may be transferred from one body to another*, according to the second law of thermodynamics.

- It is a *measurable quantity*, and so can be treated mathematically.

- It *cannot be treated as a material substance*, because it may be transformed into something that is not a material substance, e.g., mechanical work.

- Heat is *one of the forms of energy*.

From empirically based ideas of heat, and from other empirical observations, the notions of internal energy and of entropy can be derived, so as to lead to the recognition of the first and second laws of thermodynamics. This was the way of the historical pioneers of thermodynamics.

Transfers of Energy as Heat Between Two Bodies

Referring to conduction, Partington writes: "If a hot body is brought in conducting contact with a cold body, the temperature of the hot body falls and that of the cold body rises, and it is said that a *quantity of heat* has passed from the hot body to the cold body."

Referring to radiation, Maxwell writes: "In Radiation, the hotter body loses heat, and the colder body receives heat by means of a process occurring in some intervening medium which does not itself thereby become hot."

Maxwell writes that convection as such "is not a purely thermal phenomenon". In thermodynamics, convection in general is regarded as transport of internal energy. If, however, the convection is enclosed and circulatory, then it may be regarded as an intermediary that transfers energy as heat between source and destination bodies, because it transfers only energy and not matter from the source to the destination body.

Practical Operating Devices that Harness Transfers of Energy as Heat

In accordance with the first law for closed systems, energy transferred solely as heat leaves one body and enters another, changing the internal energies of each. Transfer, between bodies, of energy as work is a complementary way of changing internal energies. Though it is not logically rigorous from the viewpoint of strict physical concepts, a common form of words that expresses this is to say that heat and work are interconvertible.

Cyclically operating engines, that use only heat and work transfers, have two thermal reservoirs, a hot and a cold one. They may be classified by the range of operating temperatures of the working body, relative to those reservoirs. In a heat engine, the working body is at all times colder than the hot reservoir and hotter than the cold reservoir. In a sense, it uses heat transfer to produce work. In a heat pump, the working body, at stages of the cycle, goes both hotter than the hot reservoir, and colder than the cold reservoir. In a sense, it uses work to produce heat transfer.

Heat Engine

In classical thermodynamics, a commonly considered model is the heat engine. It consists of four bodies: the working body, the hot reservoir, the cold reservoir, and the work reservoir. A cyclic process leaves the working body in an unchanged state, and is envisaged as being repeated indefinitely often. Work transfers between the working body and the work reservoir are envisaged as reversible, and thus only one work reservoir is needed. But two thermal reservoirs are needed, because transfer of energy as heat is irreversible. A single cycle sees energy taken by the working body from the hot reservoir and sent to the two other reservoirs, the work reservoir and the cold reservoir. The hot reservoir always and only supplies energy and the cold reservoir always and only receives energy. The second law of thermodynamics requires that no cycle can occur in which no energy is received by the cold reservoir. Heat engines achieve higher efficiency when the difference between initial and final temperature is greater.

Heat Pump or Refrigerator

Another commonly considered model is the heat pump or refrigerator. Again there are four bodies: the working body, the hot reservoir, the cold reservoir, and the work reservoir. A single cycle starts with the working body colder than the cold reservoir, and then energy is taken in as heat by the working body from the cold reservoir. Then the work reservoir does work on the working body, adding more to its internal energy, making it hotter than the hot reservoir. The hot working body passes heat to the hot reservoir, but still remains hotter than the cold reservoir. Then, by allowing it to expand without doing work on another body and without passing heat to another body, the working body is made colder than the cold reservoir. It can now accept heat transfer from the cold reservoir to start another cycle.

The device has transported energy from a colder to a hotter reservoir, but this is not regarded as being by an inanimate agency; rather, it is regarded as by the harnessing of work . This is because work is supplied from the work reservoir, not just by a simple thermodynamic process, but by a cycle of thermodynamic operations and processes, which may be regarded as directed by an animate or harnessing agency. Accordingly, the cycle is still in accord with the second law of thermodynamics. The efficiency of a heat pump is best when the temperature difference between the hot and cold reservoirs is least.

Functionally, such engines are used in two ways, distinguishing a target reservoir and a resource or surrounding reservoir. A heat pump transfers heat, to the hot reservoir as the target, from the resource or surrounding reservoir. A refrigerator transfers heat, from the cold reservoir as the target, to the resource or surrounding reservoir. The target reservoir may be regarded as leaking: when the target leaks hotness to the surroundings, heat pumping is used; when the target leaks coldness to the surroundings, refrigeration is used. The engines harness work to overcome the leaks.

Macroscopic View of Quantity of Energy Transferred as Heat

According to Planck, there are three main conceptual approaches to heat. One is the microscopic or kinetic theory approach. The other two are macroscopic approaches. One is the approach through the law of conservation of energy taken as prior to thermodynamics, with a mechanical analysis of processes, for example in the work of Helmholtz. This mechanical view is taken in this article as currently customary for thermodynamic theory. The other macroscopic approach is the thermodynamic one, which admits heat as a primitive concept, which contributes, by scientific induction to knowledge of the law of conservation of energy. This view is widely taken as the practical one, quantity of heat being measured by calorimetry.

Bailyn also distinguishes the two macroscopic approaches as the mechanical and the thermodynamic. The thermodynamic view was taken by the founders of thermodynamics in the nineteenth century. It regards quantity of energy transferred as heat as a primitive concept coherent with a primitive concept of temperature, measured primarily by calorimetry. A calorimeter is a body in the surroundings of the system, with its own temperature and internal energy; when it is connected to the system by a path for heat transfer, changes in it measure heat transfer. The mechanical view was pioneered by Helmholtz and developed and used in the twentieth century, largely through the influence of Max Born. It regards quantity of heat transferred as heat as a derived concept, defined for closed systems as quantity of heat transferred by mechanisms other than work transfer, the latter being regarded as primitive for thermodynamics, defined by macroscopic mechanics. According to Born, the transfer of internal energy between open systems that accompanies transfer of matter "cannot be reduced to mechanics". It follows that there is no well-founded definition of quantities of energy transferred as heat or as work associated with transfer of matter.

Nevertheless, for the thermodynamical description of non-equilibrium processes, it is desired to consider the effect of a temperature gradient established by the surroundings across the system of interest when there is no physical barrier or wall between system and surroundings, that is to say, when they are open with respect to one another. The impossibility of a mechanical definition in terms of work for this circumstance does not alter the physical fact that a temperature gradient causes a diffusive flux of internal energy, a process that, in the thermodynamic view, might be proposed as a candidate concept for transfer of energy as heat.

In this circumstance, it may be expected that there may also be active other drivers of diffusive flux of internal energy, such as gradient of chemical potential which drives transfer of matter, and gradient of electric potential which drives electric current and iontophoresis; such effects usually interact with diffusive flux of internal energy driven by temperature gradient, and such interactions are known as cross-effects.

If cross-effects that result in diffusive transfer of internal energy were also labeled as heat transfers, they would sometimes violate the rule that pure heat transfer occurs only down a temperature gradient, never up one. They would also contradict the principle that all heat transfer is of one and the same kind, a principle founded on the idea of heat conduction between closed systems. One might to try to think narrowly of heat flux driven purely by temperature gradient as a conceptual component of diffusive internal energy flux, in the thermodynamic view, the concept resting specifically on careful calculations based on detailed knowledge of the processes and being indirectly assessed. In these circumstances, if perchance it happens that no transfer of matter is actualized, and there are no cross-effects, then the thermodynamic concept and the mechanical concept coincide, as if one were dealing with closed systems. But when there is transfer of matter, the exact laws by which temperature gradient drives diffusive flux of internal energy, rather than being exactly knowable, mostly need to be assumed, and in many cases are practically unverifiable. Consequently, when there is transfer of matter, the calculation of the pure 'heat flux' component of the diffusive flux of internal energy rests on practically unverifiable assumptions. This is a reason to think of heat as a specialized concept that relates primarily and precisely to closed systems, and applicable only in a very restricted way to open systems.

In many writings in this context, the term "heat flux" is used when what is meant is therefore more accurately called diffusive flux of internal energy; such usage of the term "heat flux" is a residue of older and now obsolete language usage that allowed that a body may have a "heat content".

Microscopic View of Heat

In the kinetic theory, heat is explained in terms of the microscopic motions and interactions of constituent particles, such as electrons, atoms, and molecules. The immediate meaning of the kinetic energy of the constituent particles is not as heat. It is as a component of internal energy. In microscopic terms, heat is a transfer quantity, and is described by a transport theory, not as steadily localized kinetic energy of particles. Heat transfer arises from temperature gradients or differences, through the diffuse exchange of microscopic kinetic and potential particle energy, by particle collisions and other interactions. An early and vague expression of this was made by Francis Bacon. Precise and detailed versions of it were developed in the nineteenth century.

In statistical mechanics, for a closed system (no transfer of matter), heat is the energy transfer associated with a disordered, microscopic action on the system, associated with jumps in occupation numbers of the energy levels of the system, without change in the values of the energy levels themselves. It is possible for macroscopic thermodynamic work to alter the occupation numbers without change in the values of the system energy levels themselves, but what distinguishes transfer as heat is that the transfer is entirely due to disordered, microscopic action, including radiative transfer. A mathematical definition can be formulated for small increments of quasi-static adiabatic work in terms of the statistical distribution of an ensemble of microstates.

Notation and Units

As a form of energy heat has the unit joule (J) in the International System of Units (SI). However, in many applied fields in engineering the British thermal unit (BTU) and the calorie are often used. The standard unit for the rate of heat transferred is the watt (W), defined as joules per second.

The total amount of energy transferred as heat is conventionally written as Q (from Quantity) for algebraic purposes. Heat released by a system into its surroundings is by convention a negative quantity ($Q < 0$); when a system absorbs heat from its surroundings, it is positive ($Q > 0$). Heat transfer rate, or heat flow per unit time, is denoted by $\dot{Q}$. This should not be confused with a time derivative of a function of state (which can also be written with the dot notation) since heat is not a function of state. Heat flux is defined as rate of heat transfer per unit cross-sectional area, resulting in the unit *watts per square metre*.

Estimation of Quantity of Heat

Quantity of heat transferred can measured by calorimetry, or determined through calculations based on other quantities.

Calorimetry is the empirical basis of the idea of quantity of heat transferred in a process. The transferred heat is measured by changes in a body of known properties, for example, temperature rise, change in volume or length, or phase change, such as melting of ice.

A calculation of quantity of heat transferred can rely on a hypothetical quantity of energy transferred as adiabatic work and on the first law of thermodynamics. Such calculation is the primary approach of many theoretical studies of quantity of heat transferred.

Latent and Sensible Heat

Joseph Black

In an 1847 lecture entitled *On Matter, Living Force, and Heat*, James Prescott Joule characterized the terms latent heat and sensible heat as components of heat each affecting distinct physical phenomena, namely the potential and kinetic energy of particles, respectively.He described latent energy as the energy possessed via a distancing of particles where attraction was over a greater distance, i.e. a form of potential energy, and the sensible heat as an energy involving the motion of particles or what was known as a *living force*. At the time of Joule kinetic energy either held 'invisibly' internally or held 'visibly' externally was known as a *living force*.

Latent heat is the heat released or absorbed by a chemical substance or a thermodynamic system during a change of state that occurs without a change in temperature.

Such a process may be a phase transition, such as the melting of ice or the boiling of water. The term was introduced around 1750 by Joseph Black as derived from the Latin *latere* (*to lie hidden*), characterizing its effect as not being directly measurable with a thermometer.

Sensible heat, in contrast to latent heat, is the heat transferred to a thermodynamic system that has as its sole effect a change of temperature.

Both latent heat and sensible heat transfers increase the internal energy of the system to which they are transferred.

Specific Heat

Specific heat, also called specific heat capacity, is defined as the amount of energy that has to be transferred to or from one unit of mass (kilogram) or amount of substance (mole) to change the system temperature by one degree. Specific heat is a physical property, which means that it depends on the substance under consideration and its state as specified by its properties.

The specific heats of monatomic gases (e.g., helium) are nearly constant with temperature. Diatomic gases such as hydrogen display some temperature dependence, and triatomic gases (e.g., carbon dioxide) still more.

Relation Between Heat and Temperature

Before the discovery of the laws of thermodynamics, quantity of energy transferred as heat was measured by changes in the states of the participating bodies.

Some general rules, with important exceptions, that will be indicated noted in following paragraphs of this section, can be stated as follows.

Most bodies, over most temperature ranges, expand on being heated. Mostly, heating a body at a constant volume increases the pressure it exerts on its constraining walls, and increases its temperature. Also mostly, heating a body at a constant pressure increases its volume, and increases its temperature.

Beyond this, most substances have three ordinarily recognized states of matter, solid, liquid, and gas, and a fourth less obviously recognized one, plasma. Many have further, more finely differentiated, states of matter, such as for example, glass, and liquid crystal. In many cases, at fixed temperature and pressure, a substance can exist in several distinct states of matter in what might be viewed as the same 'body'. For example, ice may float in a glass of water. Then the ice and the water are said to constitute two phases within the 'body'. Definite rules are known, telling how distinct phases may coexist in a 'body'. Mostly, at a fixed pressure, there is a definite temperature at which heating causes a solid to melt or evaporate, and a definite temperature at which heating causes a liquid to evaporate. In such cases, cooling has the reverse effects.

All of these, the commonest cases, fit with a rule that heating can be measured by changes of state of a body. Such cases supply what are called *thermometric bodies*, that allow the definition of empirical temperatures. Before 1848, all temperatures were defined in this way. There was thus a tight link, apparently logically determined, between heat and temperature, though they were recognized as conceptually thoroughly distinct, especially by Joseph Black in the later eighteenth century.

There are important exceptions. They break the obviously apparent link between heat and temperature. They make it clear that empirical definitions of temperature are contingent on the peculiar properties of particular thermometric substances, and are thus precluded from the title 'absolute'. For example, water contracts on being heated near 277 K. It cannot be used as a thermometric substance near that temperature. Also, over a certain temperature range, ice contracts on heating. Moreover, many substances can exist in metastable states, such as with negative pressure, that survive only transiently and in very special conditions. Such facts, sometimes called 'anomalous', are some of the reasons for the thermodynamic definition of absolute temperature.

In the early days of measurement of high temperatures, another factor was important, and used by Josiah Wedgwood in his pyrometer. The temperature reached in a process was estimated by the shrinkage of a sample of clay. The higher the temperature, the more the shrinkage. This was the only available more or less reliable method of measurement of temperatures above 1000 °C. But such shrinkage is irreversible. The clay does not expand again on cooling. That is why it could be used for the measurement. But only once. It is not a thermometric material in the usual sense of the word.

Nevertheless, the thermodynamic definition of absolute temperature does make essential use of the concept of heat, with proper circumspection.

Relation Between Hotness and Temperature

According to Denbigh, the property of hotness is a concern of thermodynamics that should be defined without reference to the concept of heat. Consideration of hotness leads to the concept of empirical temperature. All physical systems are capable of heating or cooling others. This does not require that they have thermodynamic temperatures. With reference to hotness, the comparative terms hotter and colder are defined by the rule that heat flows from the hotter body to the colder.

If a physical system is inhomogeneous or very rapidly or irregularly changing, for example by turbulence, it may be impossible to characterize it by a temperature, but still there can be transfer of energy as heat between it and another system. If a system has a physical state that is regular enough, and persists long enough to allow it to reach thermal equilibrium with a specified thermometer, then it has a temperature according to that thermometer. An empirical thermometer registers degree of hotness for such a system. Such a temperature is called empirical. For example, Truesdell writes about classical thermodynamics: "At each time, the body is assigned a real number called the *temperature*. This number is a measure of how hot the body is."

Physical systems that are too turbulent to have temperatures may still differ in hotness. A physical system that passes heat to another physical system is said to be the hotter of the two. More is required for the system to have a thermodynamic temperature. Its behavior must be so regular that its empirical temperature is the same for all suitably calibrated and scaled thermometers, and then its hotness is said to lie on the one-dimensional hotness manifold. This is part of the reason why heat is defined following Carathéodory and Born, solely as occurring other than by work or transfer of matter; temperature is advisedly and deliberately not mentioned in this now widely accepted definition.

This is also the reason why the zeroth law of thermodynamics is stated explicitly. If three physical systems, A, B, and C are each not in their own states of internal thermodynamic equilibrium, it is possible that, with suitable physical connections being made between them, A can heat B and B can heat C and C can heat A. In non-equilibrium situations, cycles of flow are possible. It is the special and uniquely distinguishing characteristic of internal thermodynamic equilibrium that this possibility is not open to thermodynamic systems (as distinguished amongst physical systems) which are in their own states of internal thermodynamic equilibrium; this is the reason why the zeroth law of thermodynamics needs explicit statement. That is to say, the relation 'is not colder than' between general non-equilibrium physical systems is not transitive, whereas, in contrast, the relation 'has no lower a temperature than' between thermodynamic systems in their own states of internal thermodynamic equilibrium is transitive. It follows from this that the relation 'is in thermal equilibrium with' is transitive, which is one way of stating the zeroth law.

Just as temperature may undefined for a sufficiently inhomogeneous system, so also may entropy be undefined for a system not in its own state of internal thermodynamic equilibrium. For example, 'the temperature of the solar system' is not a defined quantity. Likewise, 'the entropy of the solar system' is not defined in classical thermodynamics. It has not been possible to define non-equilibrium entropy, as a simple number for a whole system, in a clearly satisfactory way.

Rigorous Definition of Quantity of Energy Transferred as Heat

It is sometimes convenient to have a rigorous definition of quantity of energy transferred as heat. Such a definition is customarily based on the work of Carathéodory (1909), referring to processes in a closed system, as follows.

The internal energy U_X of a body in an arbitrary state X can be determined by amounts of work adiabatically performed by the body on its surrounds when it starts from a reference state O. Such work is assessed through quantities defined in the surroundings of the body. It is supposed that such work can be assessed accurately, without error due to friction in the surroundings; friction in the body is not excluded by this definition. The adiabatic performance of work is defined in terms of adiabatic walls, which allow transfer of energy as work, but no other transfer, of energy or matter. In particular they do not allow the passage of energy as heat. According to this definition, work performed adiabatically is in general accompanied by friction within the thermodynamic system or body. On the other hand, according to Carathéodory (1909), there also exist non-adiabatic walls, which are postulated to be "permeable only to heat", and are called diathermal.

For the definition of quantity of energy transferred as heat, it is customarily envisaged that an arbitrary state of interest Y is reached from state O by a process with two components, one adiabatic and the other not adiabatic. For convenience one may say that the adiabatic component was the sum of work done by the body through volume change through movement of the walls while the non-adiabatic wall was temporarily rendered adiabatic, and of isochoric adiabatic work. Then the non-adiabatic component is a process of energy transfer through the wall that passes only heat, newly made accessible for the purpose of this transfer, from the surroundings to the body. The change in internal energy to reach the state Y from the state O is the difference of the two amounts of energy transferred.

Although Carathéodory himself did not state such a definition, following his work it is customary in theoretical studies to define the quantity of energy transferred as heat, Q, to the body from its surroundings, in the combined process of change to state Y from the state O, as the change in internal energy, ΔU_Y, minus the amount of work, W, done by the body on its surrounds by the adiabatic process, so that $Q = \Delta U_Y - W$.

In this definition, for the sake of conceptual rigour, the quantity of energy transferred as heat is not specified directly in terms of the non-adiabatic process. It is defined through knowledge of precisely two variables, the change of internal energy and the amount of adiabatic work done, for the combined process of change from the reference state O to the arbitrary state Y. It is important that this does not explicitly involve the amount of energy transferred in the non-adiabatic component of the combined process. It is assumed here that the amount of energy required to pass from state O to state Y, the change of internal energy, is known, independently of the combined process, by a determination through a purely adiabatic process, like that for the determination of the internal energy of state X above. The rigour that is prized in this definition is that there is one and only one kind of energy transfer admitted as fundamental: energy transferred as work. Energy transfer as heat is considered as a derived quantity. The uniqueness of work in this scheme is considered to guarantee rigor and purity of conception. The conceptual purity of this definition, based on the concept of energy transferred as work as an ideal notion, relies on the idea that some frictionless and otherwise non-dissipative processes of energy transfer can be realized in physical actuality. The second law of thermodynamics, on the other hand, assures us that such processes are not found in nature.

Heat, Temperature, and Thermal Equilibrium Regarded as Jointly Primitive Notions

Before the rigorous mathematical definition of heat based on Carathéodory's 1909 paper, recounted just above, historically, heat, temperature, and thermal equilibrium were presented in thermodynamics textbooks as jointly primitive notions. Carathéodory introduced his 1909 paper thus: "The proposition that the discipline of thermodynamics can be justified without recourse to any hypothesis that cannot be verified experimentally must be regarded as one of the most noteworthy results of the research in thermodynamics that was accomplished during the last century." Referring to the "point of view adopted by most authors who were active in the last fifty years", Carathéodory wrote: "There exists a physical quantity called heat that is not identical with the mechanical quantities (mass, force, pressure, etc.) and whose variations can be determined by calorimetric measurements." James Serrin introduces an account of the theory of thermodynamics thus: "In the following section, we shall use the classical notions of *heat*, *work*, and *hotness* as primitive elements, ... That heat is an appropriate and natural primitive for thermodynamics was already accepted by Carnot. Its continued validity as a primitive element of thermodynamical structure is due to the fact that it synthesizes an essential physical concept, as well as to its successful use in recent work to unify different constitutive theories." This traditional kind of presentation of the basis of thermodynamics includes ideas that may be summarized by the statement that heat transfer is purely due to spatial non-uniformity of temperature, and is by conduction and radiation, from hotter to colder bodies. It is sometimes proposed that this traditional kind of presentation necessarily rests on "circular reasoning"; against this proposal, there stands the rigorously logical mathematical development of the theory presented by Truesdell and Bharatha (1977).

This alternative approach to the definition of quantity of energy transferred as heat differs in logical structure from that of Carathéodory, recounted just above.

This alternative approach admits calorimetry as a primary or direct way to measure quantity of energy transferred as heat. It relies on temperature as one of its primitive concepts, and used in calorimetry. It is presupposed that enough processes exist physically to allow measurement of differences in internal energies. Such processes are not restricted to adiabatic transfers of energy as work. They include calorimetry, which is the commonest practical way of finding internal energy differences. The needed temperature can be either empirical or absolute thermodynamic.

In contrast, the Carathéodory way recounted just above does not use calorimetry or temperature in its primary definition of quantity of energy transferred as heat.

The Carathéodory way regards calorimetry only as a secondary or indirect way of measuring quantity of energy transferred as heat. As recounted in more detail just above, the Carathéodory way regards quantity of energy transferred as heat in a process as primarily or directly defined as a residual quantity. It is calculated from the difference of the internal energies of the initial and final states of the system, and from the actual work done by the system during the process. That internal energy difference is supposed to have been measured in advance through processes of purely adiabatic transfer of energy as work, processes that take the system between the initial and final states. By the Carathéodory way it is presupposed as known from experiment that there actually physically exist enough such adiabatic processes, so that there need be no recourse to calorimetry for measurement of quantity of energy transferred as heat. This presupposition is essential but is explicitly labeled neither as a law of thermodynamics nor as an axiom of the Carathéodory way. In fact, the actual physical existence of such adiabatic processes is indeed mostly supposition, and those supposed processes have in most cases not been actually verified empirically to exist.

Heat Transfer in Engineering

The discipline of heat transfer, typically considered an aspect of mechanical engineering and chemical engineering, deals with specific applied methods by which thermal energy in a system is generated, or converted, or transferred to another system. Although the definition of heat implicitly means the transfer of energy, the term *heat transfer* encompasses this traditional usage in many engineering disciplines and laymen language.

Heat transfer includes the mechanisms of heat conduction, thermal radiation, and mass transfer.

In engineering, the term *convective heat transfer* is used to describe the combined effects of conduction and fluid flow. From the thermodynamic point of view, heat flows into a fluid by diffusion to increase its energy, the fluid then transfers (advects) this increased internal energy (not heat) from one location to another, and this is then followed by a second thermal interaction which transfers heat to a second body or system, again by diffusion. This entire process is often regarded as an additional mechanism of heat transfer, although technically, "heat transfer" and thus heating and cooling occurs only on either end of such a conductive flow, but not as a result of flow. Thus, conduction can be said to "transfer" heat only as a net result of the process, but may not do so at every time within the complicated convective process.

Although distinct physical laws may describe the behavior of each of these methods, real systems often exhibit a complicated combination which are often described by a variety of mathematical methods.

Heat Capacity

Heat capacity or thermal capacity is a measurable physical quantity equal to the ratio of the heat added to (or removed from) an object to the resulting temperature change. The unit of heat capacity is joule per kelvin $\frac{J}{K}$, or kilogram metre squared per kelvin second squared $\frac{kgm^2}{Ks^2}$ in the International System of Units (SI). The dimensional form is $L^2MT^{-2}\Theta^{-1}$. Specific heat is the amount of heat needed to raise the temperature of one kilogram of mass by 1 kelvin.

Heat capacity is an extensive property of matter, meaning it is proportional to the size of the system. When expressing the same phenomenon as an intensive property, the heat capacity is divided by the amount of substance, mass, or volume, thus the quantity is independent of the size or extent of the sample. The molar heat capacity is the heat capacity per unit amount (SI unit: mole) of a pure substance and the specific heat capacity, often simply called specific heat, is the heat capacity per unit mass of a material. Nonetheless some authors state the term specific heat to refer to the ratio of the specific heat capacity of a substance at any given temperature, to the specific heat capacity of another substance at a reference temperature, much in the fashion of a specific gravity. Occasionally, in engineering contexts, the volumetric heat capacity is used.

Temperature reflects the average randomized kinetic energy of constituent particles of matter (e.g. atoms or molecules) relative to the centre of mass of the system, while heat is the transfer of energy across a system boundary into the body other than by work or matter transfer. Translation, rotation, and vibration of atoms represent the degrees of freedom of motion which classically contribute to the heat capacity of gases, while only vibrations are needed to describe the heat capacities of most solids , as shown by the Dulong–Petit law. Other contributions can come from magnetic and electronic degrees of freedom in solids, but these rarely make substantial contributions.

For quantum mechanical reasons, at any given temperature, some of these degrees of freedom may be unavailable, or only partially available, to store thermal energy. In such cases, the specific heat capacity is a fraction of the maximum. As the temperature approaches absolute zero, the specific capacity of a system approaches zero, due to loss of available degrees of freedom. Quantum theory can be used to quantitatively predict the specific heat capacity of simple systems.

History

In a previous theory of heat common in the early modern period, heat was thought to be a measurement of an invisible fluid, known as the *caloric*. Bodies were capable of holding a certain amount of this fluid, hence the term *heat capacity*, named and first investigated by Scottish chemist Joseph Black in the 1750s.

Since the development of thermodynamics during the 18th and 19th centuries, scientists have abandoned the idea of a physical caloric, and instead understand heat as changes in a system's internal energy. That is, heat is no longer considered a fluid; rather, heat is a transfer of disordered energy. Nevertheless, at least in English, the term "heat capacity" survives. In some other languages, the term *thermal capacity* is preferred, and it is also sometimes used in English.

Units

Extensive Properties

In the International System of Units, heat capacity has the unit joules per kelvin (J/K). An object's heat capacity (symbol C) is defined as the ratio of the amount of heat energy transferred to an object and the resulting increase in temperature of the object,

$$C = \frac{Q}{\Delta T},$$

assuming that the temperature range is sufficiently small thus the heat capacity is constant. More generally, because heat capacity does depend upon temperature, it should be written as

$$C(T) = \frac{\delta Q}{dT},$$

where the symbol δ is used to imply that heat is a path function. Heat capacity is an extensive property, meaning it depends on the extent or size of the physical system in question. A sample containing twice the amount of substance as another sample requires the transfer of twice the amount of heat (Q) to achieve the same change in temperature (ΔT).

Alternative Unit Systems

While SI units are the most widely used, some countries and industries also use other systems of measurement. One older unit of heat is the kilogram-calorie (Cal), originally defined as the energy required to raise the temperature of one kilogram of water by one degree Celsius, typically from 14.5 to 15.5 °C. The specific average heat capacity of water on this scale would therefore be exactly 1 Cal/(C°·kg). However, due to the temperature-dependence of the specific heat, a large number of different definitions of the calorie came into being. Whilst once it was very prevalent, especially its smaller cgs variant the gram-calorie (cal), defined thus the specific heat of water would be 1 cal/(K·g), in most fields the use of the calorie is now archaic.

In the United States other units of measure for heat capacity may be quoted in disciplines such as construction, civil engineering, and chemical engineering. A still common system is the English Engineering Units in which the mass reference is pound mass and the temperature is specified in degrees Fahrenheit or Rankine. One (rare) unit of heat is the pound calorie (lb-cal), defined as the amount of heat required to raise the temperature of one pound of water by one degree Celsius. On this scale the specific heat of water would be 1 lb-cal/(K·lbm). More common is the British thermal unit, the standard unit of heat in the U.S. construction industry. This is defined such that the specific heat of water is 1 BTU/(F°·lb). The path integral Monte Carlo method is a numerical approach for determining the values of heat capacity, based on quantum dynamical principles. However, good approximations can be made for gases in many states using simpler methods outlined below. For many solids composed of relatively heavy atoms (atomic number > iron), at non-cryogenic temperatures, the heat capacity at room temperature approaches 3R = 24.94 joules per kelvin per mole of atoms (Dulong–Petit law, R is the gas constant). Low temperature approximations for both gases and solids at temperatures less than their characteristic Einstein temperatures or Debye temperatures can be made by the methods of Einstein and Debye discussed below. Water (liquid): CP = 4185.5 J/(kg·K) (15 °C, 101.325 kPa) Water (liquid): CVH = 74.539 J/(mol·K) (25 °C) For liquids and gases, it is important to know the pressure to which given heat capacity data refer. Most published data are given for standard pressure. However, different standard conditions for temperature and pressure have been defined by different organizations. The International Union of Pure and Applied Chemistry (IUPAC) changed its recommendation from one atmosphere to the round value 100 kPa ($\approx$750.062 Torr).

Measurement of Heat Capacity

It may appear that the way to measure heat capacity is to add a known amount of heat to an object, and measure the change in temperature. This works reasonably well for many solids. However, for precise measurements, and especially for gases, other aspects of measurement become critical.

The heat capacity can be affected by many of the state variables that describe the thermodynamic system under study. These include the starting and ending temperature, as well as the pressure and the volume of the system before and after heat is added. So rather than a single way to measure heat capacity, there are actually several slightly different measurements of heat capacity. The most commonly used methods for measurement are to hold the object either at constant pressure (C_p) or at constant volume (C_v). Gases and liquids are typically also measured at constant volume. Measurements under constant pressure produce larger values than those at constant volume because the constant pressure values also include heat energy that is used to do work to expand the substance against the constant pressure as its temperature increases. This difference is particularly notable in gases where values under constant pressure are typically 30% to 66.7% greater than those at constant volume. Hence the heat capacity ratio of gases is typically between 1.3 and 1.67.

The specific heat capacities of substances comprising molecules (as distinct from monatomic gases) are not fixed constants and vary somewhat depending on temperature. Accordingly, the temperature at which the measurement is made is usually also specified. Examples of two common ways to cite the specific heat of a substance are as follows:

- Water (liquid): C_p = 4185.5 J/(kg·K) (15 °C, 101.325 kPa)

- Water (liquid): C_vH = 74.539 J/(mol·K) (25 °C)

For liquids and gases, it is important to know the pressure to which given heat capacity data refer. Most published data are given for standard pressure. However, quite different standard conditions for temperature and pressure have been defined by different organizations. The International Union of Pure and Applied Chemistry (IUPAC) changed its recommendation from one atmosphere to the round value 100 kPa ($\approx$750.062 Torr).

Theory of Heat Capacity

Factors that Affect Specific Heat Capacity

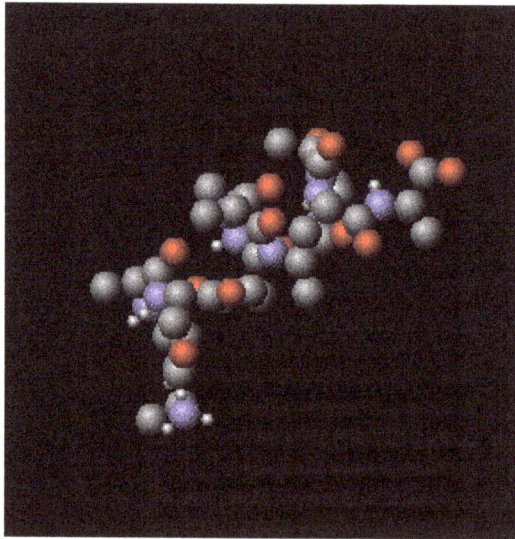

Molecules undergo many characteristic internal vibrations. Potential energy stored in these internal degrees of freedom contributes to a sample's energy content,   but not to its temperature. More internal degrees of freedom tend to increase a substance's specific heat capacity, so long as temperatures are high enough to overcome quantum effects.

For any given substance, the heat capacity of a body is directly proportional to the amount of substance it contains (measured in terms of mass or moles or volume). Doubling the amount of substance in a body doubles its heat capacity, etc.

However, when this effect has been corrected for, by dividing the heat capacity by the quantity of substance in a body, the resulting specific heat capacity is a function of the structure of the substance itself. In particular, it depends on the number of degrees of freedom that are available to the particles in the substance; each independent degree of freedom allows the particles to store thermal energy. The translational kinetic energy of substance particles which manifests as *temperature change* is only one of the many possible degrees of freedom, and thus the larger the number of degrees of freedom available to the particles of a substance *other* than translational kinetic energy, the larger will be the specific heat capacity for the substance. For example, rotational kinetic energy of gas molecules stores heat energy in a way that increases heat capacity, since this energy does not contribute to temperature.

In addition, quantum effects require that whenever energy be stored in any mechanism associated with a bound system which confers a degree of freedom, it must be stored in certain minimal-sized deposits (quanta) of energy, or else not stored at all. Such effects limit the full ability of some degrees of freedom to store energy when their lowest energy storage quantum amount is not easily supplied at the average energy of particles at a given temperature. In general, for this reason, specific heat capacities tend to fall at lower temperatures where the average thermal energy available to each particle degree of freedom is smaller, and thermal energy storage begins to be limited by these quantum effects. Due to this process, as temperature falls toward absolute zero, so also does heat capacity.

Degrees of Freedom

Molecules are quite different from the monatomic gases like helium and argon. With monatomic gases, thermal energy comprises only translational motions. Translational motions are ordinary, whole-body movements in 3D space whereby particles move about and exchange energy in collisions—like rubber balls in a vigorously shaken container. These simple movements in the three dimensions of space mean individual atoms have three translational degrees of freedom. A degree of freedom is any form of energy in which heat transferred into an object can be stored. This can be in translational kinetic energy, rotational kinetic energy, or other forms such as potential energy in vibrational modes. Only three translational degrees of freedom (corresponding to the three independent directions in space) are available for any individual atom, whether it is free, as a monatomic molecule, or bound into a polyatomic molecule.

As to rotation about an atom's axis (again, whether the atom is bound or free), its energy of rotation is proportional to the moment of inertia for the atom, which is extremely small compared to moments of inertia of collections of atoms. This is because almost all of the mass of a single atom is concentrated in its nucleus, which has a radius too small to give a significant moment of inertia. In contrast, the *spacing* of quantum energy levels for a rotating object is inversely proportional to its moment of inertia, and so this spacing becomes very large for objects with very small moments of inertia. For these reasons, the contribution from rotation of atoms on their axes is essentially zero in monatomic gases, because the energy spacing of the associated quantum levels is too large for significant thermal energy to be stored in rotation of systems with such small moments of inertia. For similar reasons, axial rotation around bonds joining atoms in diatomic gases (or along the linear axis in a linear molecule of any length) can also be neglected as a possible "degree of freedom" as well, since such rotation is similar to rotation of monatomic atoms, and so occurs about an axis with a moment of inertia too small to be able to store significant heat energy.

In polyatomic molecules, other rotational modes may become active, due to the much higher moments of inertia about certain axes which do not coincide with the linear axis of a linear molecule. These modes take the place of some translational degrees of freedom for individual atoms, since the atoms are moving in 3-D space, as the molecule rotates. The narrowing of quantum mechanically determined energy spacing between rotational states results from situations where atoms are rotating around an axis that does not connect them, and thus form an assembly that has a large moment of inertia. This small difference between energy states allows the kinetic energy of this type of rotational motion to store heat energy at ambient temperatures. Furthermore, internal vibrational degrees of freedom also may become active (these are also a type of translation, as seen from the view of each atom). In summary, molecules are complex objects with a population of atoms that may move about within the molecule in a number of different ways, and each of these ways of moving is capable of storing energy if the temperature is sufficient.

The heat capacity of molecular substances (on a "per-atom" or atom-molar, basis) does not exceed the heat capacity of monatomic gases, unless vibrational modes are brought into play. The reason for this is that vibrational modes allow energy to be stored as potential energy in inter-atomic bonds in a molecule, which are not available to atoms in monatomic gases. Up to about twice as much energy (on a per-atom basis) per unit of temperature increase can be stored in a solid as in a monatomic gas, by this mechanism of storing energy in the potentials of interatomic bonds. This gives many solids about twice the atom-molar heat capacity at room temperature of monatomic gases.

However, quantum effects heavily affect the actual ratio at lower temperatures (i.e., much lower than the melting temperature of the solid), especially in solids with light and tightly bound atoms (e.g., beryllium metal or diamond). Polyatomic gases store intermediate amounts of energy, giving them a "per-atom" heat capacity that is between that of monatomic gases ($\frac{3}{2} R$ per mole of atoms, where R is the ideal gas constant), and the maximum of fully excited warmer solids ($3 R$ per mole of atoms). For gases, heat capacity never falls below the minimum of $\frac{3}{2} R$ per mole (of molecules), since the kinetic energy of gas molecules is always available to store at least this much thermal energy. However, at cryogenic temperatures in solids, heat capacity falls toward zero, as temperature approaches absolute zero.

Example of Temperature-dependent Specific Heat Capacity, in a Diatomic Gas

To illustrate the role of various degrees of freedom in storing heat, we may consider nitrogen, a diatomic molecule that has five active degrees of freedom at room temperature: the three comprising translational motions plus two rotational degrees of freedom internally. Although the constant-volume molar heat capacity of nitrogen at this temperature is five-thirds that of monatomic gases, on a per-mole of atoms basis, it is five-sixths that of a monatomic gas. The reason for this is the loss of a degree of freedom due to the bond when it does not allow storage of thermal energy. Two separate nitrogen atoms would have a total of six degrees of freedom—the three translational degrees of freedom of each atom. When the atoms are bonded the molecule will still only have three translational degrees of freedom, as the two atoms in the molecule move as one. However, the molecule cannot be treated as a point object, and the moment of inertia has increased sufficiently about two axes to allow two rotational degrees of freedom to be active at room temperature to give five degrees of freedom. The moment of inertia about the third axis remains small, as this is the axis passing through the centres of the two atoms, and so is similar to the small moment of inertia for atoms of a monatomic gas. Thus, this degree of freedom does not act to store heat, and does not contribute to the heat capacity of nitrogen. The heat capacity *per atom* for nitrogen (5/2 R per mole molecules = 5/4 R per mole atoms) is therefore less than for a monatomic gas (3/2 R per mole molecules or atoms), so long as the temperature remains low enough that no vibrational degrees of freedom are activated.

At higher temperatures, however, nitrogen gas gains one more degree of internal freedom, as the molecule is excited into higher vibrational modes that store thermal energy. A vibrational degree of freedom contributes a heat capacity of 1/2 R each for kinetic and potential energy, for a total of R. Now the bond is contributing heat capacity, and (because of storage of energy in potential energy) is contributing more than if the atoms were not bonded. With full thermal excitation of bond vibration, the heat capacity per volume, or per mole of gas *molecules* approaches seven-thirds that of monatomic gases. Significantly, this is seven-sixths of the monatomic gas value on a mole-of-atoms basis, so this is now a *higher* heat capacity *per atom* than the monatomic figure, because the vibrational mode enables for diatomic gases allows an extra degree of *potential energy* freedom per pair of atoms, which monatomic gases cannot possess.

However, even at these large temperatures where gaseous nitrogen is able to store 7/6[ths] of the energy *per atom* of a monatomic gas (making it more efficient at storing energy on an atomic basis), it still only stores 7/12 [ths] of the maximal per-atom heat capacity of a *solid,* meaning it is not nearly as efficient at storing thermal energy on an atomic basis, as solid substances can be. This is typical of gases, and results because many of the potential bonds which might be storing potential energy in gaseous nitrogen (as opposed to solid nitrogen) are lacking, because only one of the spatial dimensions for each nitrogen atom offers a bond into which potential energy can be stored without increasing the kinetic energy of the atom. In general, solids are most efficient, on an atomic basis, at storing thermal energy (that is, they have the highest per-atom or per-mole-of-atoms heat capacity).

Per Mole of Different Units

Per mole of molecules

When the specific heat capacity, c, of a material is measured (lowercase c means the unit quantity is in terms of mass), different values arise because different substances have different molar masses (essentially, the weight of the individual atoms or molecules). In solids, thermal energy arises due to the number of atoms that are vibrating. "Molar" heat capacity *per mole of molecules*, for both gases and solids, offer figures which are arbitrarily large, since molecules may be arbitrarily large. Such heat capacities are thus not intensive quantities for this reason, since the quantity of mass being considered can be increased without limit.

Per mole of atoms

Conversely, for *molecular-based* substances (which also absorb heat into their internal degrees of freedom), massive, complex molecules with high atomic count—like octane—can store a great deal of energy per mole and yet are quite unremarkable on a mass basis, or on a per-atom basis. This is because, in fully excited systems, heat is stored independently by each atom in a substance, not primarily by the bulk motion of molecules.

Thus, it is the heat capacity per-mole-of-atoms, not per-mole-of-molecules, which is the intensive quantity, and which comes closest to being a constant for all substances at high temperatures. This relationship was noticed empirically in 1819, and is called the Dulong–Petit law, after its two discoverers. Historically, the fact that specific heat capacities are approximately equal when corrected by the presumed weight of the atoms of solids, was an important piece of data in favor of the atomic theory of matter.

Because of the connection of heat capacity to the number of atoms, some care should be taken to specify a mole-of-molecules basis vs. a mole-of-atoms basis, when comparing specific heat capacities of molecular solids and gases. Ideal gases have the same numbers of molecules per volume, so increasing molecular complexity adds heat capacity on a per-volume and per-mole-of-molecules basis, but may lower or raise heat capacity on a per-atom basis, depending on whether the temperature is sufficient to store energy as atomic vibration.

In solids, the quantitative limit of heat capacity in general is about $3R$ per mole of atoms, where R is the ideal gas constant. This $3R$ value is about 24.9 J/mole.K. Six degrees of freedom (three kinetic and three potential) are available to each atom. Each of these six contributes $\frac{1}{2}R$ specific heat capacity per mole of atoms. This limit of $3R$ per mole specific heat capacity is approached at room temperature for most solids, with significant departures at this temperature only for solids composed of the lightest atoms which are bound very strongly, such as beryllium (where the value is only of 66% of 3 R), or diamond (where it is only 24% of 3 R). These large departures are due to quantum effects which prevent full distribution of heat into all vibrational modes, when the energy difference between vibrational quantum states is very large compared to the average energy available to each atom from the ambient temperature.

For monatomic gases, the specific heat is only half of 3 R per mole, i.e. ($\frac{3}{2}$R per mole) due to loss of all potential energy degrees of freedom in these gases. For polyatomic gases, the heat capacity will be intermediate between these values on a per-mole-of-atoms basis, and (for heat-stable molecules) would approach the limit of 3 R per mole of atoms, for gases composed of complex molecules, and at higher temperatures at which all vibrational modes accept excitational energy. This is because very large and complex gas molecules may be thought of as relatively large blocks of solid matter which have lost only a relatively small fraction of degrees of freedom, as compared to a fully integrated solid.

Corollaries of these Considerations for Solids (Volume-specific Heat Capacity)

Since the bulk density of a solid chemical element is strongly related to its molar mass (usually about 3 R per mole, as noted above), there exists a noticeable inverse correlation between a solid's density and its specific heat capacity on a per-mass basis. This is due to a very approximate tendency of atoms of most elements to be about the same size (and constancy of mole-specific heat capacity) resulting in a good correlation between the *volume* of any given solid chemical element and its total heat capacity. Another way of stating this, is that the volume-specific heat capacity (volumetric heat capacity) of solid elements is roughly a constant. The molar volume of solid elements is very roughly constant, and (even more reliably) so also is the molar heat capacity for most solid substances. These two factors determine the volumetric heat capacity, which as a bulk property may be striking in consistency. For example, the element uranium is a metal which has a density almost 36 times that of the metal lithium, but uranium's specific heat capacity on a volumetric basis (i.e. per given volume of metal) is only 18% larger than lithium's.

Since the volume-specific corollary of the Dulong–Petit specific heat capacity relationship requires that atoms of all elements take up (on average) the same volume in solids, there are many departures from it, with most of these due to variations in atomic size. For instance, arsenic, which is only 14.5% less dense than antimony, has nearly 59% more specific heat capacity on a mass basis. In other words; even though an ingot of arsenic is only about 17% larger than an antimony one of the same mass, it absorbs about 59% more heat for a given temperature rise. The heat capacity ratios of the two substances closely follows the ratios of their molar volumes (the ratios of numbers of atoms in the same volume of each substance); the departure from the correlation to simple volumes in this case is due to lighter arsenic atoms being significantly more closely packed than antimony atoms, instead of similar size. In other words, similar-sized atoms would cause a mole of arsenic to be 63% larger than a mole of antimony, with a correspondingly lower density, allowing its volume to more closely mirror its heat capacity behavior.

Other Factors

Hydrogen bonds

Hydrogen-containing polar molecules like ethanol, ammonia, and water have powerful, intermolecular hydrogen bonds when in their liquid phase. These bonds provide another place where heat may be stored as potential energy of vibration, even at comparatively low temperatures. Hydrogen bonds account for the fact that liquid water stores nearly the theoretical limit of $3R$ per mole of atoms, even at relatively low temperatures (i.e. near the freezing point of water).

Impurities

In the case of alloys, there are several conditions in which small impurity concentrations can greatly affect the specific heat. Alloys may exhibit marked difference in behaviour even in the case of small amounts of impurities being one element of the alloy; for example impurities in semiconducting ferromagnetic alloys may lead to quite different specific heat properties.

The constant-pressure heat capacity for any gas would exceed this by an extra factor of R. As example C_p would be a total of $(15/2)R/mole$ for nitrous oxide.

The Effect of Quantum Energy Levels in Storing Energy in Degrees of Freedom

The various degrees of freedom cannot generally be considered to obey classical mechanics, however. Classically, the energy residing in each degree of freedom is assumed to be continuous—it can take on any positive value, depending on the temperature. In reality, the amount of energy that may reside in a particular degree of freedom is quantized: It may only be increased and decreased in finite amounts. A good estimate of the size of this minimum amount is the energy of the first excited state of that degree of freedom above its ground state. For example, the first vibrational state of the hydrogen chloride (HCl) molecule has an energy of about 5.74×10^{-20} joule. If this amount of energy were deposited in a classical degree of freedom, it would correspond to a temperature of about 4156 K.

If the temperature of the substance is so low that the equipartition energy of $(1/2)kT$ is much smaller than this excitation energy, then there will be little or no energy in this degree of freedom. This degree of freedom is then said to be "frozen out". As mentioned above, the temperature corresponding to the first excited vibrational state of HCl is about 4156 K. For temperatures well below this value, the vibrational degrees of freedom of the HCl molecule will be frozen out. They will contain little energy and will not contribute to the thermal energy or the heat capacity of HCl gas.

Energy Storage Mode "Freeze-out" Temperatures

It can be seen that for each degree of freedom there is a critical temperature at which the degree of freedom "unfreezes" and begins to accept energy in a classical way. In the case of translational degrees of freedom, this temperature is that temperature at which the thermal wavelength of the molecules is roughly equal to the size of the container. For a container of macroscopic size (e.g. 10 cm) this temperature is extremely small and has no significance, since the gas will certainly liquify or freeze before this low temperature is reached. For any real gas translational degrees of freedom may be considered to always be classical and contain an average energy of $(3/2)kT$ per molecule.

The rotational degrees of freedom are the next to "unfreeze". In a diatomic gas, for example, the critical temperature for this transition is usually a few tens of kelvins, although with a very light molecule such as hydrogen the rotational energy levels will be spaced so widely that rotational heat capacity may not completely "unfreeze" until considerably higher temperatures are reached. Finally, the vibrational degrees of freedom are generally the last to unfreeze. As an example, for diatomic gases, the critical temperature for the vibrational motion is usually a few thousands of kelvins, and thus for the nitrogen in our example at room temperature, no vibration modes would be excited, and the constant-volume heat capacity at room temperature is $(5/2)R$/mole, not $(7/2)R$/mole. As seen above, with some unusually heavy gases such as iodine gas I_2, or bromine gas Br_2, some vibrational heat capacity may be observed even at room temperatures.

It should be noted that it has been assumed that atoms have no rotational or internal degrees of freedom. This is in fact untrue. For example, atomic electrons can exist in excited states and even the atomic nucleus can have excited states as well. Each of these internal degrees of freedom are assumed to be frozen out due to their relatively high excitation energy. Nevertheless, for sufficiently high temperatures, these degrees of freedom cannot be ignored. In a few exceptional cases, such molecular electronic transitions are of sufficiently low energy that they contribute to heat capacity at room temperature, or even at cryogenic temperatures. One example of an electronic transition degree of freedom which contributes heat capacity at standard temperature is that of nitric oxide (NO), in which the single electron in an anti-bonding molecular orbital has energy transitions which contribute to the heat capacity of the gas even at room temperature.

An example of a nuclear magnetic transition degree of freedom which is of importance to heat capacity, is the transition which converts the spin isomers of hydrogen gas (H_2)

into each other. At room temperature, the proton spins of hydrogen gas are aligned 75% of the time, resulting in *orthohydrogen* when they are. Thus, some thermal energy has been stored in the degree of freedom available when *parahydrogen* (in which spins are anti-aligned) absorbs energy, and is converted to the higher energy ortho form. However, at the temperature of liquid hydrogen, not enough heat energy is available to produce orthohydrogen (that is, the transition energy between forms is large enough to "freeze out" at this low temperature), and thus the parahydrogen form predominates. The heat capacity of the transition is sufficient to release enough heat, as orthohydrogen converts to the lower-energy parahydrogen, to boil the hydrogen liquid to gas again, if this evolved heat is not removed with a catalyst after the gas has been cooled and condensed. This example also illustrates the fact that some modes of storage of heat may not be in con-stant equilibrium with each other in substances, and heat absorbed or released from such phase changes may "catch up" with temperature changes of substances, only after a cer-tain time. In other words, the heat evolved and absorbed from the ortho-para isomeric transition contributes to the heat capacity of hydrogen on long time-scales, but not on *short* time-scales. These time scales may also depend on the presence of a catalyst.

Less exotic phase-changes may contribute to the heat-capacity of substances and systems, as well, as (for example) when water is converted back and forth from solid to liquid or gas form. Phase changes store heat energy entirely in breaking the bonds of the potential ener-gy interactions between molecules of a substance. As in the case of hydrogen, it is also pos-sible for phase changes to be hindered as the temperature drops, thus they do not catch up and become apparent, without a catalyst. For example, it is possible to supercool liquid wa-ter to below the freezing point, and not observe the heat evolved when the water changes to ice, so long as the water remains liquid. This heat appears instantly when the water freezes.

Solid Phase

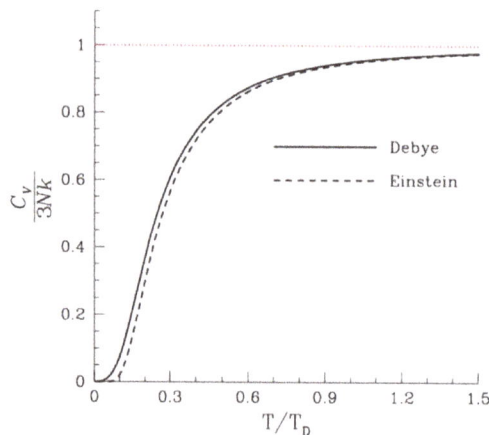

The dimensionless heat capacity divided by three, as a function of temperature as predicted by the Debye model and by Einstein's earlier model. The horizontal axis is the temperature divided by the Debye temperature. Note that, as expected, the dimensionless heat capacity is zero at absolute zero, and rises to a value of three as the temperature becomes much larger than the Debye temperature. The red line corresponds to the classical limit of the Dulong–Petit law

For matter in a crystalline solid phase, the Dulong–Petit law, which was discovered empirically, states that the molar heat capacity assumes the value 3 R. Indeed, for solid metallic chemical elements at room temperature, molar heat capacities range from about 2.8 R to 3.4 R. Large exceptions at the lower end involve solids composed of relatively low-mass, tightly bonded atoms, such as beryllium at 2.0 R, and diamond at only 0.735 R. The latter conditions create larger quantum vibrational energy spacing, thus many vibrational modes have energies too high to be populated (and thus are "frozen out") at room temperature. At the higher end of possible heat capacities, heat capacity may exceed R by modest amounts, due to contributions from anharmonic vibrations in solids, and sometimes a modest contribution from conduction electrons in metals. These are not degrees of freedom treated in the Einstein or Debye theories.

The theoretical maximum heat capacity for multi-atomic gases at higher temperatures, as the molecules become larger, also approaches the Dulong–Petit limit of 3 R, so long as this is calculated per mole of atoms, not molecules. The reason for this behavior is that, in theory, gases with very large molecules have almost the same high-temperature heat capacity as solids, lacking only the (small) heat capacity contribution that comes from potential energy that cannot be stored between separate molecules in a gas.

The Dulong–Petit limit results from the equipartition theorem, and as such is only valid in the classical limit of a microstate continuum, which is a high temperature limit. For light and non-metallic elements, as well as most of the common molecular solids based on carbon compounds at standard ambient temperature, quantum effects may also play an important role, as they do in multi-atomic gases. These effects usually combine to give heat capacities lower than 3 R per mole of *atoms* in the solid, although in molecular solids, heat capacities calculated *per mole of molecules* in molecular solids may be more than 3 R. For example, the heat capacity of water ice at the melting point is about 4.6 R per mole of molecules, but only 1.5 R per mole of atoms. As noted, heat capacity values far lower than 3 R "per atom" (as is the case with diamond and beryllium) result from "freezing out" of possible vibration modes for light atoms at suitably low temperatures, just as happens in many low-mass-atom gases at room temperatures (where vibrational modes are all frozen out). Because of high crystal binding energies, the effects of vibrational mode freezing are observed in solids more often than liquids: for example the heat capacity of liquid water is twice that of ice at near the same temperature, and is again close to the 3 R per mole of atoms of the Dulong–Petit theoretical maximum.

Liquid Phase

A general theory of the heat capacity of liquids has not yet been achieved, and is still an active area of research. It was long thought that phonon theory is not able to explain the heat capacity of liquids, since liquids only sustain longitudinal, but not transverse phonons, which in solids are responsible for 2/3 of the heat capacity. However, Brillouin scattering experiments with neutrons and with X-rays, confirming an intuition of

Yakov Frenkel, have shown that transverse phonons do exist in liquids, albeit restricted to frequencies above a threshold called the Frenkel frequency. Since most energy is contained in these high-frequency modes, a simple modification of the Debye model is sufficient to yield a good approximation to experimental heat capacities of simple liquids.

Amorphous materials can be considered a type of liquid. The specific heat of amorphous materials has characteristic discontinuities at the glass transition temperature. These discontinuities are frequently used to detect the glass transition temperature where a supercooled liquid transforms to a glass.

Table of Specific Heat Capacities

Note that the especially high molar values, as for paraffin, gasoline, water and ammonia, result from calculating specific heats in terms of moles of *molecules*. If specific heat is expressed per mole of *atoms* for these substances, none of the constant-volume values exceed, to any large extent, the theoretical Dulong–Petit limit of 25 $J \cdot mol^{-1} \cdot K^{-1}$ = 3 R per mole of atoms. Paraffin, for example, has very large molecules and thus a high heat capacity per mole, but as a substance it does not have remarkable heat capacity in terms of volume, mass, or atom-mol (which is just 1.41 R per mole of atoms, or less than half of most solids, in terms of heat capacity per atom).

In the last column, major departures of solids at standard temperatures from the Dulong–Petit law value of 3 R, are usually due to low atomic weight plus high bond strength (as in diamond) causing some vibration modes to have too much energy to be available to store thermal energy at the measured temperature. For gases, departure from 3 R per mole of atoms in this table is generally due to two factors: **(1)** failure of the higher quantum-energy-spaced vibration modes in gas molecules to be excited at room temperature, and **(2)** loss of potential energy degree of freedom for small gas molecules, simply because most of their atoms are not bonded maximally in space to other atoms, as happens in many solids.

Notable minima and maxima are shown in maroon						
Table of specific heat capacities at 25 °C (298 K) unless otherwise noted. Substance	Phase	Isobaric mass heat capacity c_p $J \cdot g^{-1} \cdot K^{-1}$	Isobaricmolar heat capacity $C_{P,m}$ $J \cdot mol^{-1} \cdot K^{-1}$	Isochore molar heat capacity $C_{V,m}$ $J \cdot mol^{-1} \cdot K^{-1}$	Isobaric volumetric heat capacity $C_{P,v}$ $J \cdot cm^{-3} \cdot K^{-1}$	Isochore atom-molar heat capacity in units of R $C_{V,am}$ atom-mol^{-1}
Air (Sea level, dry, 0 °C (273.15 K))	gas	1.0035	29.07	20.7643	0.001297	~ 1.25 R

Air (typical room conditions[A])	gas	1.012	29.19	20.85	0.00121	~ 1.25 R
Aluminium	solid	0.897	24.2		2.422	2.91 R
Ammonia	liquid	4.700	80.08		3.263	3.21 R
Animal tissue (incl. human)	mixed	3.5			3.7*	
Antimony	solid	0.207	25.2		1.386	3.03 R
Argon	gas	0.5203	20.7862	12.4717		1.50 R
Arsenic	solid	0.328	24.6		1.878	2.96 R
Beryllium	solid	1.82	16.4		3.367	1.97 R
Bismuth	solid	0.123	25.7		1.20	3.09 R
Cadmium	solid	0.231	26.02			3.13 R
Carbon dioxide CO_2	gas	0.839*	36.94	28.46		1.14 R
Chromium	solid	0.449	23.35			2.81 R
Copper	solid	0.385	24.47		3.45	2.94 R
Diamond	solid	0.5091	6.115		1.782	0.74 R
Ethanol	liquid	2.44	112		1.925	1.50 R
Gasoline (octane)	liquid	2.22	228		1.64	1.05 R
Glass	solid	0.84			2.1	
Gold	solid	0.129	25.42		2.492	3.05 R
Granite	solid	0.790			2.17	
Graphite	solid	0.710	8.53		1.534	1.03 R
Helium	gas	5.1932	20.7862	12.4717		1.50 R
Hydrogen	gas	14.30	28.82			1.23 R
Hydrogen sulfide H_2S	gas	1.015*	34.60			1.05 R
Iron	solid	0.412	25.09		3.537	3.02 R
Lead	solid	0.129	26.4		1.44	3.18 R
Lithium	solid	3.58	24.8		1.912	2.98 R
Lithium at 181 °C	liquid	4.379	30.33		2.242	3.65 R
Magnesium	solid	1.02	24.9		1.773	2.99 R
Mercury	liquid	0.1395	27.98		1.888	3.36 R
Methane at 2 °C	gas	2.191	35.69			0.85 R
Methanol	liquid	2.14	68.62			1.38 R
Molten salt (142–540 °C)	liquid	1.56			2.62	
Nitrogen	gas	1.040	29.12	20.8		1.25 R
Neon	gas	1.0301	20.7862	12.4717		1.50 R
Oxygen	gas	0.918	29.38	21.0		1.26 R

Substance	Phase	Isobaric mass heat capacity c_p $J \cdot g^{-1} \cdot K^{-1}$	Isobaric molar heat capacity $C_{P,m}$ $J \cdot mol^{-1} \cdot K^{-1}$	Isochore molar heat capacity $C_{V,m}$ $J \cdot mol^{-1} \cdot K^{-1}$	Isobaric volumetric heat capacity $C_{P,v}$ $J \cdot cm^{-3} \cdot K^{-1}$	Isochore atom-molar heat capacity in units of R $C_{V,am}$ atom-mol^{-1}
Paraffin wax $C_{25}H_{52}$	solid	2.5 (ave)	900		2.325	1.41 R
Polyethylene (rotomolding grade)	solid	2.3027				
Silica (fused)	solid	0.703	42.2		1.547	1.69 R
Silver	solid	0.233	24.9		2.44	2.99 R
Sodium	solid	1.230	28.23			3.39 R
Steel	solid	0.466			3.756	
Tin	solid	0.227	27.112		1.659	3.26 R
Titanium	solid	0.523	26.060		2.6384	3.13 R
Tungsten	solid	0.134	24.8		2.58	2.98 R
Uranium	solid	0.116	27.7		2.216	3.33 R
Water at 100 °C (steam)	gas	2.080	37.47	28.03		1.12 R
Water at 25 °C	liquid	4.1813	75.327	74.53	4.1796	3.02 R
Water at 100 °C	liquid	4.1813	75.327	74.53	4.2160	3.02 R
Water at −10 °C (ice)	solid	2.05	38.09		1.938	1.53 R
Zinc	solid	0.387	25.2		2.76	3.03 R

A Assuming an altitude of 194 metres above mean sea level (the world–wide median altitude of human habitation), an indoor temperature of 23 °C, a dewpoint of 9 °C (40.85% relative humidity), and 760 mm–Hg sea level–corrected barometric pressure (molar water vapor content = 1.16%).*Derived data by calculation. This is for water-rich tissues such as brain. The whole-body average figure for mammals is approximately 2.9 $J \cdot cm^{-3} \cdot K^{-1}$

Mass Heat Capacity of Building Materials

(Usually of interest to builders and solar designers)

Mass heat capacity of building materials		
Substance	Phase	c_p $J \cdot g^{-1} \cdot K^{-1}$
Asphalt	solid	0.920
Brick	solid	0.840
Concrete	solid	0.880
Glass, silica	solid	0.840

Substance	Phase	c_p J g^{-1} K^{-1}
Glass, crown	solid	0.670
Glass, flint	solid	0.503
Glass, pyrex	solid	0.753
Granite	solid	0.790
Gypsum	solid	1.090
Marble, mica	solid	0.880
Sand	solid	0.835
Soil	solid	0.800
Water	liquid	4.1813
Wood	solid	1.7 (1.2 to 2.9)

Latent Heat

Latent heat is energy released or absorbed, by a body or a thermodynamic system, during a constant-temperature process. An example is latent heat of fusion for a phase change, melting, at a specified temperature and pressure. The term was introduced around 1762 by Scottish chemist Joseph Black. It is derived from the Latin *latere* (*to lie hidden*). Black used the term in the context of calorimetry where a heat transfer caused a volume change while the thermodynamic system's temperature was constant.

In contrast to latent heat, sensible heat involves an energy transfer that results in a temperature change of the system.

Usage

The terms "sensible heat" and "latent heat" are specific forms of energy; they are two properties of a material or in a thermodynamic system. "Sensible heat" is a body's internal energy that may be "sensed" or felt. "Latent heat" is internal energy concerning the phase (solid / liquid / gas) of a material and does not affect the temperature.

Both sensible and latent heats are observed in many processes of transport of energy in nature. changes of Latent heat is associated with the change of phase of atmospheric water, vaporization and condensation, whereas sensible heat is energy that reflects the temperature of the atmosphere or ocean, or ice.

The original usage of the term, as introduced by Black, was applied to systems that were intentionally held at constant temperature. Such usage referred to *latent heat of expansion* and several other related latent heats. These latent heats are defined independently of the conceptual framework of thermodynamics.

When a body is heated at constant temperature by thermal radiation in a microwave field for example, it may expand by an amount described by its *latent heat with respect to volume* or *latent heat of expansion*, or increase its pressure by an amount described by its *latent heat with respect to pressure*.

Two common forms of latent heat are latent heat of fusion (melting) and latent heat of vaporization (boiling). These names describe the direction of energy flow when changing from one phase to the next: from solid to liquid, and liquid to gas.

In both cases the change is endothermic, meaning that the system absorbs energy. For example, when water evaporates, energy is required for the water molecules to overcome the forces of attraction between them, the transition from water to vapor requires an input of energy.

If the vapor then condenses to a liquid on a surface, then the vapor's latent energy absorbed during evaporation is released as the liquid's sensible heat onto the surface.

The large value of the enthalpy of condensation of water vapor is the reason that steam is a far more effective heating medium than boiling water, and is more hazardous.

Meteorology

In meteorology, latent heat flux is the flux of heat from the Earth's surface to the atmosphere that is associated with evaporation or transpiration of water at the surface and subsequent condensation of water vapor in the troposphere. It is an important component of Earth's surface energy budget. Latent heat flux has been commonly measured with the Bowen ratio technique, or more recently since the mid-1900s by the eddy covariance method.

History

The English word *latent* comes from Latin *latēns*, meaning *lying hidden*. The term *latent heat* was introduced into calorimetry around 1750 when Joseph Black, commissioned by producers of Scotch whisky in search of ideal quantities of fuel and water for their distilling process, to studying system changes, such as of volume and pressure, when the thermodynamic system was held at constant temperature in a thermal bath. James Prescott Joule characterised latent energy as the energy of interaction in a given configuration of particles, i.e. a form of potential energy, and the sensible heat as an energy that was indicated by the thermometer, relating the latter to thermal energy.

Specific Latent Heat

A *specific* latent heat (L) expresses the amount of energy in the form of heat (Q) required to completely effect a phase change of a unit of mass (m), usually 1kg, of a substance as an intensive property:

$$L = \frac{Q}{m}.$$

Intensive properties are material characteristics and are not dependent on the size or extent of the sample. Commonly quoted and tabulated in the literature are the specific latent heat of fusion and the specific latent heat of vaporization for many substances.

From this definition, the latent heat for a given mass of a substance is calculated by

$$Q = mL$$

where:

> Q is the amount of energy released or absorbed during the change of phase of the substance (in kJ or in BTU),
>
> m is the mass of the substance (in kg or in lb), and
>
> L is the specific latent heat for a particular substance (kJ-kg$_m^{-1}$ or in BTU-lb$_m^{-1}$), either L_f for fusion, or L_v for vaporization.

Table of Specific Latent Heats

The following table shows the specific latent heats and change of phase temperatures of some common fluids and gases.

Substance	S.L.H. of Fusion kJ/kg	Melting Point °C	S.L.H. of Vapor-ization kJ/kg	Boiling Point °C
Alcohol, ethyl	108	−114	855	78.3
Ammonia	332.17	−77.74	1369	−33.34
Carbon dioxide	184	−78	574	−57
Helium			21	−268.93
Hydrogen(2)	58	−259	455	−253
Lead	23.0	327.5	871	1750
Nitrogen	25.7	−210	200	−196
Oxygen	13.9	−219	213	−183
Refrigerant R134a		−101	215.9	−26.6
Refrigerant R152a		−116	326.5	-25
Toluene	72.1	−93	351	110.6
Turpentine			293	
Water	334	0	2264.76	100

Specific Latent Heat for Condensation of Water

The specific latent heat of condensation of water in the temperature range from −25 °C to 40 °C is approximated by the following empirical cubic function:

$$L_{water}(T) = (2500.8 - 2.36T + 0.0016T^2 - 0.00006T^3)\,\text{J/g},$$

where the temperature T is taken to be the numerical value in °C.

For sublimation and deposition from and into ice, the specific latent heat is almost constant in the temperature range from −40 °C to 0 °C and can be approximated by the following empirical quadratic function:

$$L_{ice}(T) = (2834.1 - 0.29T - 0.004T^2)\,\text{J/g}.$$

Temperature

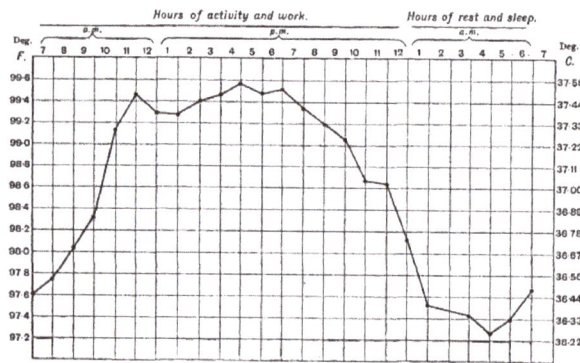

Body temperature variation

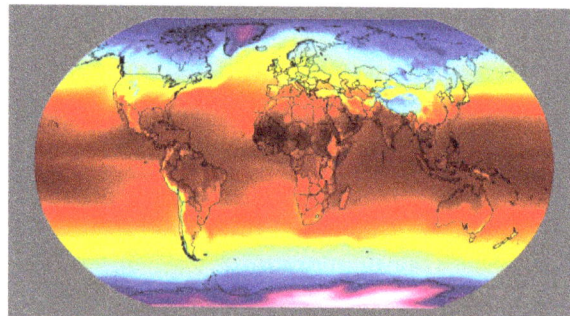

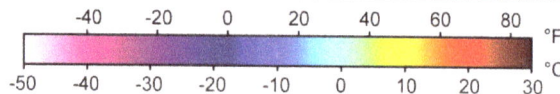

Annual Mean Temperature

Annual mean temperature around the world

A temperature is an objective comparative measurement of hot or cold. It is measured by a thermometer. Several scales and units exist for measuring temperature, the most common being Celsius (denoted °C; formerly called *centigrade*), Fahrenheit (denoted °F), and, especially in science, Kelvin (denoted K).

The coldest theoretical temperature is absolute zero, at which the thermal motion of atoms and molecules reaches its minimum – classically, this would be a state of motionlessness, but quantum uncertainty dictates that the particles still possess a finite zero-point energy. Absolute zero is denoted as 0 K on the Kelvin scale, −273.15 °C on the Celsius scale, and −459.67 °F on the Fahrenheit scale.

The kinetic theory offers a valuable but limited account of the behavior of the materials of macroscopic bodies, especially of fluids. It indicates the absolute temperature as proportional to the average kinetic energy of the random microscopic motions of those of their constituent microscopic particles, such as electrons, atoms, and molecules, that move freely within the material.

Temperature is important in all fields of natural science, including physics, geology, chemistry, atmospheric sciences, medicine, and biology, as well as most aspects of daily life.

Effects of Temperature

Many physical processes are affected by temperature, such as

- physical properties of materials including the phase (solid, liquid, gaseous or plasma), density, solubility, vapor pressure, electrical conductivity

- rate and extent to which chemical reactions occur

- the amount and properties of thermal radiation emitted from the surface of an object

- speed of sound is a function of the square root of the absolute temperature

Temperature Scales

Temperature scales differ in two ways: the point chosen as zero degrees, and the magnitudes of incremental units or degrees on the scale.

The Celsius scale (°C) is used for common temperature measurements in most of the world. It is an empirical scale. It developed by a historical progress, which led to its zero point 0°C being defined by the freezing point of water, with additional degrees defined so that 100°C was the boiling point of water, both at sea-level atmospheric pressure. Because of the 100 degree interval, it is called a centigrade scale. Since the standardization of the kelvin in the International System of Units, it has subsequently been redefined in terms of the equivalent fixing points on the Kelvin scale, and so that a temperature increment of one degree Celsius is the same as an increment of one kelvin, though they differ by an additive offset of 273.15.

The United States commonly uses the Fahrenheit scale, on which water freezes at 32°F and boils at 212°F at sea-level atmospheric pressure.

Many scientific measurements use the Kelvin temperature scale (unit symbol: K), named in honor of the Scottish physicist who first defined it. It is a thermodynamic or absolute temperature scale. Its zero point, 0K, is defined to coincide with the coldest physically-possible temperature (called absolute zero). Its degrees are defined through thermodynamics. The temperature of absolute zero occurs at 0K = −273.15°C (or −459.67°F), and the freezing point of water at sea-level atmospheric pressure occurs at 273.15K = 0°C.

The International System of Units (SI) defines a scale and unit for the kelvin or thermodynamic temperature by using the reliably reproducible temperature of the triple point of water as a second reference point (the first reference point being 0 K at absolute zero). The triple point is a singular state with its own unique and invariant temperature and pressure, along with, for a fixed mass of water in a vessel of fixed volume, an autonomically and stably self-determining partition into three mutually contacting phases, vapour, liquid, and solid, dynamically depending only on the total internal energy of the mass of water. For historical reasons, the triple point temperature of water is fixed at 273.16 units of the measurement increment.

Thermodynamic Approach to Temperature

Temperature is one of the principal quantities in the study of thermodynamics.

Kinds of Temperature Scale

There is a variety of kinds of temperature scale. It may be convenient to classify them as empirically and theoretically based. Empirical temperature scales are historically older, while theoretically based scales arose in the middle of the nineteenth century.

Empirically Based Scales

Empirically based temperature scales rely directly on measurements of simple physical properties of materials. For example, the length of a column of mercury, confined in a glass-walled capillary tube, is dependent largely on temperature, and is the basis of the very useful mercury-in-glass thermometer. Such scales are valid only within convenient ranges of temperature. For example, above the boiling point of mercury, a mercury-in-glass thermometer is impracticable. Most materials expand with temperature increase, but some materials, such as water, contract with temperature increase over some specific range, and then they are hardly useful as thermometric materials. A material is of no use as a thermometer near one of its phase-change temperatures, for example its boiling-point.

In spite of these restrictions, most generally used practical thermometers are of the empirically based kind. Especially, it was used for calorimetry, which contributed greatly to the discovery of thermodynamics. Nevertheless, empirical thermometry has serious drawbacks when judged as a basis for theoretical physics. Empirically based thermometers, beyond their base as simple direct measurements of ordinary physical properties of thermometric materials, can be re-calibrated, by use of theoretical physical reasoning, and this can extend their range of adequacy.

Theoretically Based Scales

Theoretically based temperature scales are based directly on theoretical arguments, especially those of thermodynamics, of kinetic theory, and of quantum mechanics. They rely on theoretical properties of idealized devices and materials. They are more or less comparable with practically feasible physical devices and materials. Theoretically based temperature scales are used to provide calibrating standards for practical empirically based thermometers.

The accepted fundamental thermodynamic temperature scale is the Kelvin scale, based on an ideal cyclic process envisaged for a Carnot heat engine.

An ideal material on which a temperature scale can be based is the ideal gas. The pressure exerted by a fixed volume and mass of an ideal gas is directly proportional to its temperature. Some natural gases show so nearly ideal properties over suitable temperature ranges that they can be used for thermometry; this was important during the development of thermodynamics, and is still of practical importance today. The ideal gas thermometer is, however, not theoretically perfect for thermodynamics. This is because the entropy of an ideal gas at its absolute zero of temperature is not a positive semi-definite quantity, which puts the gas in violation of the third law of thermodynamics. The physical reason is that the ideal gas law, exactly read, refers to the limit of infinitely high temperature and zero pressure.

Measurement of the spectrum of electromagnetic radiation from an ideal three-dimensional black body can provide an accurate temperature measurement because the frequency of maximum spectral radiance of black-body radiation is directly proportional to the temperature of the black body; this is known as Wien's displacement law, and has a theoretical explanation in Planck's law and the Bose–Einstein law.

Measurement of the spectrum of noise-power produced by an electrical resistor can also provide an accurate temperature measurement. The resistor has two terminals and is in effect a one-dimensional body. The Bose-Einstein law for this case indicates that the noise-power is directly proportional to the temperature of the resistor and to the value of its resistance and to the noise band-width. In a given frequency band, the noise-power has equal contributions from every frequency, and is called Johnson noise. If the value of the resistance is known then the temperature can be found.

If molecules, or atoms, or electrons, are emitted from a material and their velocities are measured, the spectrum of their velocities often nearly obeys a theoretical law called the Maxwell–Boltzmann distribution, which gives a well-founded measurement of temperatures for which the law holds. There have not yet been successful experiments of this same kind that directly use the Fermi–Dirac distribution for thermometry, but perhaps that will be achieved in future.

Absolute Thermodynamic Scale

The Kelvin scale is called absolute for two reasons. One is Kelvin's, that its formal character is independent of the properties of particular materials. The other reason is that its zero is in a sense absolute, in that it indicates absence of microscopic classical motion of the constituent particles of matter, so that they have a limiting specific heat of zero for zero temperature, according to the third law of thermodynamics. Nevertheless, a Kelvin temperature has a definite numerical value, that has been arbitrarily chosen by tradition. This numerical value also depends on the properties of water, which has a gas–liquid–solid triple point that can be reliably reproduced as a standard experimental phenomenon. The choice of this triple point is also arbitrary and by convention. The Kelvin scale is also called the thermodynamic scale.

Definition of the Kelvin Scale

The thermodynamic definition of temperature is due to Kelvin.

It is framed in terms of an idealized device called a Carnot engine, imagined to define a continuous cycle of states of its working body. The cycle is imagined to run so slowly that at each point of the cycle the working body is in a state of thermodynamic equilibrium. There are four limbs in such a Carnot cycle. The engine consists of four bodies. The main one is called the working body. Two of them are called heat reservoirs, so large that their respective non-deformation variables are not changed by transfer of energy as heat through a wall permeable only to heat to the working body. The fourth body is able to exchange energy with the working body only through adiabatic work; it may be called the work reservoir. The substances and states of the two heat reservoirs should be chosen so that they are not in thermal equilibrium with one another. This means that they must be at different fixed temperatures, one, labeled here with the number 1, hotter than the other, labeled here with the number 2. This can be tested by connecting the heat reservoirs successively to an auxiliary empirical thermometric body that starts each time at a convenient fixed intermediate temperature. The thermometric body should be composed of a material that has a strictly monotonic relation between its chosen empirical thermometric variable and the amount of adiabatic isochoric work done on it. In order to settle the structure and sense of operation of the Carnot cycle, it is convenient to use such a material also for the working body; because most materials are of this kind, this is hardly a restriction of the generality of this definition. The Carnot cycle is considered to start from an initial condition of the working body that was reached by the completion of a reversible adiabatic compression. From there, the working body is initially connected by a wall permeable only to heat to the heat reservoir number 1, so that during the first limb of the cycle it expands and does work on the work reservoir. The second limb of the cycle sees the working body expand adiabatically and reversibly, with no energy exchanged as heat, but more energy being transferred as work to the work reservoir. The third limb of the cycle sees the working body connected, through a wall permeable only to heat, to the heat reservoir 2, contracting and accepting energy as work from the work reservoir. The cycle is closed by reversible adiabatic compression of the working body, with no energy transferred as heat, but energy being transferred to it as work from the work reservoir.

With this set-up, the four limbs of the reversible Carnot cycle are characterized by amounts of energy transferred, as work from the working body to the work reservoir, and as heat from the heat reservoirs to the working body. The amounts of energy transferred as heat from the heat reservoirs are measured through the changes in the non-deformation variable of the working body, with reference to the previously known properties of that body, the amounts of work done on the work reservoir, and the first law of thermodynamics. The amounts of energy transferred as heat respectively from reservoir 1 and from reservoir 2 may then be denoted respectively Q_1 and Q_2. Then the absolute or thermodynamic temperatures, T_1 and T_2, of the reservoirs are defined so that to be such that

$$T_1 / T_2 = -Q_1 / Q_2. \tag{1}$$

Kelvin's original work postulating absolute temperature was published in 1848. It was based on the work of Carnot, before the formulation of the first law of thermodynamics. Kelvin wrote in his 1848 paper that his scale was absolute in the sense that it was defined "independently of the properties of any particular kind of matter." His definitive publication, which sets out the definition just stated, was printed in 1853, a paper read in 1851.

This definition rests on the physical assumption that there are readily available walls permeable only to heat. In his detailed definition of a wall permeable only to heat, Carathéodory includes several ideas. The non-deformation state variable of a closed system is represented as a real number. A state of thermal equilibrium between two closed systems connected by a wall permeable only to heat means that a certain mathematical relation holds between the state variables, including the respective non-deformation variables, of those two systems (that particular mathematical relation is regarded by Buchdahl as a preferred statement of the zeroth law of thermodynamics). Also, referring to thermal contact equilibrium, "whenever each of the systems S_1 and S_2 is made to reach equilibrium with a third system S_3 under identical conditions, the systems S_1 and S_2 are in mutual equilibrium." It may be viewed as a re-statement of the principle stated by Maxwell in the words: "All heat is of the same kind." This physical idea is also expressed by Bailyn as a possible version of the zeroth law of thermodynamics: "All diathermal walls are equivalent." Thus the present definition of thermodynamic temperature rests on the zeroth law of thermodynamics. Explicitly, this present definition of thermodynamic temperature also rests on the first law of thermodynamics, for the determination of amounts of energy transferred as heat.

Implicitly for this definition, the second law of thermodynamics provides information that establishes the virtuous character of the temperature so defined. It provides that any working substance that complies with the requirement stated in this definition will lead to the same ratio of thermodynamic temperatures, which in this sense is universal, or absolute. The second law of thermodynamics also provides that the thermodynamic temperature defined in this way is positive, because this definition requires that the heat reservoirs not be in thermal equilibrium with one another, and the cycle can be imagined to operate only in one sense if net work is to be supplied to the work reservoir.

Numerical details are settled by making one of the heat reservoirs a cell at the triple point of water, which is defined to have an absolute temperature of 273.16 K. The zeroth law of thermodynamics allows this definition to be used to measure the absolute or thermodynamic temperature of an arbitrary body of interest, by making the other heat reservoir have the same temperature as the body of interest.

Temperature as an Intensive Variable

In thermodynamic terms, temperature is an intensive variable because it is equal to a differential coefficient of one extensive variable with respect to another, for a given body. It thus has the dimensions of a ratio of two extensive variables. In thermodynamics, two bodies are often considered as connected by contact with a common wall, which has some specific permeability properties. Such specific permeability can be referred to a specific intensive variable. An example is a diathermic wall that is permeable only to heat; the intensive variable for this case is temperature. When the two bodies have been in contact for a very long time, and have settled to a permanent steady state, the relevant intensive variables are equal in the two bodies; for a diathermal wall, this statement is sometimes called the zeroth law of thermodynamics.

In particular, when the body is described by stating its internal energy U, an extensive variable, as a function of its entropy S, also an extensive variable, and other state variables V, N, with $U = U(S, V, N)$, then the temperature is equal to the partial derivative of the internal energy with respect to the entropy:

$$T = \left(\frac{\partial U}{\partial S} \right)_{V,N}. \qquad (2)$$

Likewise, when the body is described by stating its entropy S as a function of its internal energy U, and other state variables V, N, with $S = S(U, V, N)$, then the reciprocal of the temperature is equal to the partial derivative of the entropy with respect to the internal energy:

$$\frac{1}{T} = \left(\frac{\partial S}{\partial U} \right)_{V,N}. \tag{3}$$

The above definition, equation (1), of the absolute temperature is due to Kelvin. It refers to systems closed to transfer of matter, and has special emphasis on directly experimental procedures. A presentation of thermodynamics by Gibbs starts at a more abstract level and deals with systems open to the transfer of matter; in this development of thermodynamics, the equations (2) and (3) above are actually alternative definitions of temperature.

Temperature Local when Local Thermodynamic Equilibrium Prevails

Real world bodies are often not in thermodynamic equilibrium and not homogeneous. For study by methods of classical irreversible thermodynamics, a body is usually spatially and temporally divided conceptually into 'cells' of small size. If classical thermodynamic equilibrium conditions for matter are fulfilled to good approximation in such a 'cell', then it is homogeneous and a temperature exists for it. If this is so for every 'cell' of the body, then local thermodynamic equilibrium is said to prevail throughout the body.

It makes good sense, for example, to say of the extensive variable U, or of the extensive variable S, that it has a density per unit volume, or a quantity per unit mass of the system, but it makes no sense to speak of density of temperature per unit volume or quantity of temperature per unit mass of the system. On the other hand, it makes no sense to speak of the internal energy at a point, while when local thermodynamic equilibrium prevails, it makes good sense to speak of the temperature at a point. Consequently, temperature can vary from point to point in a medium that is not in global thermodynamic equilibrium, but in which there is local thermodynamic equilibrium.

Thus, when local thermodynamic equilibrium prevails in a body, temperature can be regarded as a spatially varying local property in that body, and this is because temperature is an intensive variable.

Kinetic Theory Approach to Temperature

A more thorough account of this is below at Theoretical foundation.

Kinetic theory provides a microscopic explanation of temperature, based on macroscopic systems' being composed of many microscopic particles, such as molecules and ions of various species, the particles of a species being all alike. It explains macroscopic phenomena through the classical mechanics of the microscopic particles. The equipartition theorem of kinetic theory asserts that each classical degree of freedom of a freely moving particle has an average kinetic energy of $k_B T/2$ where k_B denotes Boltzmann's constant. The translational motion of the particle has three degrees of freedom, so that, except at very low temperatures where quantum effects predominate, the average translational kinetic energy of a freely moving particle in a system with temperature T will be $3k_BT/2$.

It is possible to measure the average kinetic energy of constituent microscopic particles if they are allowed to escape from the bulk of the system. The spectrum of velocities has to be measured, and the average calculated from that. It is not necessarily the case that the particles that escape and are measured have the same velocity distribution as the particles that remain in the bulk of the system, but sometimes a good sample is possible.

Molecules, such as oxygen (O_2), have more degrees of freedom than single spherical atoms: they undergo rotational and vibrational motions as well as translations. Heating results in an increase in temperature due to an increase in the average translational kinetic energy of the molecules. Heating will also cause, through equipartitioning, the energy associated with vibrational and rotational modes to increase. Thus a diatomic gas will require more energy input to increase its temperature by a certain amount, i.e. it will have a greater heat capacity than a monatomic gas.

The process of cooling involves removing internal energy from a system. When no more energy can be removed, the system is at absolute zero, though this cannot be achieved experimentally. Absolute zero is the null point of the thermodynamic temperature scale, also called absolute temperature. If it were possible to cool a system to absolute zero, all classical motion of its particles would cease and they would be at complete rest in this classical sense. Microscopically in the description of quantum mechanics, however, matter still has zero-point energy even at absolute zero, because of the uncertainty principle.

Basic Theory

Temperature is a measure of a quality of a state of a material The quality may be regarded as a more abstract entity than any particular temperature scale that measures it, and is called *hotness* by some writers. The quality of hotness refers to the state of material only in a particular locality, and in general, apart from bodies held in a steady state of thermodynamic equilibrium, hotness varies from place to place. It is not necessarily the case that a material in a particular place is in a state that is steady and nearly homogeneous enough to allow it to have a well-defined hotness or temperature. Hotness may be represented abstractly as a one-dimensional manifold. Every valid temperature scale has its own one-to-one map into the hotness manifold.

When two systems in thermal contact are at the same temperature no heat transfers between them. When a temperature difference does exist heat flows spontaneously from the warmer system to the colder system until they are in thermal equilibrium. Heat transfer occurs by conduction or by thermal radiation.

Experimental physicists, for example Galileo and Newton, found that there are indefinitely many empirical temperature scales. Nevertheless, the zeroth law of thermodynamics says that they all measure the same quality.

Temperature for Bodies in Thermodynamic Equilibrium

For experimental physics, hotness means that, when comparing any two given bodies in their respective separate thermodynamic equilibria, any two suitably given empirical thermometers with numerical scale readings will agree as to which is the hotter of the two given bodies, or that they have the same temperature. This does not require the two thermometers to have a linear relation between their numerical scale readings, but it does require that the relation between their numerical readings shall be strictly monotonic. A definite sense of greater hotness can be had, independently of calorimetry, of thermodynamics, and of properties of particular materials, from Wien's displacement law of thermal radiation: the temperature of a bath of thermal radiation is proportional, by a universal constant, to the frequency of the maximum of its frequency spectrum; this frequency is always positive, but can have values that tend to zero. Thermal radiation is initially defined for a cavity in thermodynamic equilibrium. These physical facts justify a mathematical statement that hotness exists on an ordered one-dimensional manifold. This is a fundamental character of temperature and thermometers for bodies in their own thermodynamic equilibrium.

Except for a system undergoing a first-order phase change such as the melting of ice, as a closed system receives heat, without change in its volume and without change in external force fields acting on it, its temperature rises. For a system undergoing such a phase change so slowly that departure from thermodynamic equilibrium can be neglected, its temperature remains constant as the system is supplied with latent heat. Conversely, a loss of heat from a closed system, without phase change, without change of volume, and without change in external force fields acting on it, decreases its temperature.

Temperature for Bodies in a Steady State but not in Thermodynamic Equilibrium

While for bodies in their own thermodynamic equilibrium states, the notion of temperature requires that all empirical thermometers must agree as to which of two bodies is the hotter or that they are at the same temperature, this requirement is not safe for bodies that are in steady states though not in thermodynamic equilibrium. It can then well be that different empirical thermometers disagree about which is the hotter, and if this is so, then at least one of the bodies does not have a well defined absolute thermodynamic temperature. Nevertheless, any one given body and any one suitable empirical thermometer can still support notions of empirical, non-absolute, hotness and temperature, for a suitable range of processes. This is a matter for study in non-equilibrium thermodynamics.

Temperature for Bodies not in a Steady State

When a body is not in a steady state, then the notion of temperature becomes even less safe than for a body in a steady state not in thermodynamic equilibrium. This is also a matter for study in non-equilibrium thermodynamics.

Thermodynamic Equilibrium Axiomatics

For axiomatic treatment of thermodynamic equilibrium, since the 1930s, it has become customary to refer to a zeroth law of thermodynamics. The customarily stated minimalist version of such a law postulates only that all bodies, which when thermally connected would be in thermal equilibrium, should be said to have the same temperature by definition, but by itself does not establish temperature as a quantity expressed as a real number on a scale. A more physically informative version of such a law views empirical temperature as a chart on a hotness manifold. While the zeroth law permits the definitions of many different empirical scales of temperature, the second law of thermodynamics selects the definition of a single preferred, absolute temperature, unique up to an arbitrary scale factor, whence called the thermodynamic temperature. If internal energy is considered as a function of the volume and entropy of a homogeneous system in thermodynamic equilibrium, thermodynamic absolute temperature appears as the partial derivative of internal energy with respect the entropy at constant volume. Its natural, intrinsic origin or null point is absolute zero at which the entropy of any system is at a minimum. Although this is the lowest absolute temperature described by the model, the third law of thermodynamics postulates that absolute zero cannot be attained by any physical system.

Heat Capacity

When an energy transfer to or from a body is only as heat, state of the body changes. Depending on the surroundings and the walls separating them from the body, various changes are possible in the body. They include chemical reactions, increase of pressure, increase of temperature, and phase change. For each kind of change under specified conditions, the heat capacity is the ratio of the quantity of heat transferred to the magnitude of the change. For example, if the change is an increase in temperature at constant volume, with no phase change and no chemical change, then the temperature of the body rises and its pressure increases. The quantity of heat transferred, ΔQ, divided by the observed temperature change, ΔT, is the body's heat capacity at constant volume, C_V.

$$C_V = \frac{\Delta Q}{\Delta T}$$

If heat capacity is measured for a well defined amount of substance, the specific heat is the measure of the heat required to increase the temperature of such a unit quantity by one unit of temperature. For example, to raise the temperature of water by one kelvin (equal to one degree Celsius) requires 4186 joules per kilogram (J/kg)..

Temperature Measurement

Temperature measurement using modern scientific thermometers and temperature scales goes back at least as far as the early 18th century, when Gabriel Fahrenheit adapted a thermometer (switching to mercury) and a scale both developed by Ole Christensen Rømer. Fahrenheit's scale is still in use in the United States for non-scientific applications.

A typical Celsius thermometer measures a winter day temperature of −17°C

Temperature is measured with thermometers that may be calibrated to a variety of temperature scales. In most of the world (except for Belize, Myanmar, Liberia and the United States), the Celsius scale is used for most temperature measuring purposes. Most scientists measure temperature using the Celsius scale and thermodynamic temperature using the Kelvin scale, which is the Celsius scale offset so that its null point is 0K = −273.15°C, or absolute zero. Many engineering fields in the U.S., notably high-tech and US federal specifications (civil and military), also use the Kelvin and Celsius scales. Other engineering fields in the U.S. also rely upon the Rankine scale (a shifted Fahrenheit scale) when working in thermodynamic-related disciplines such as combustion.

Units

The basic unit of temperature in the International System of Units (SI) is the kelvin. It has the symbol K.

For everyday applications, it is often convenient to use the Celsius scale, in which 0°C corresponds very closely to the freezing point of water and 100°C is its boiling point at sea level. Because liquid droplets commonly exist in clouds at sub-zero temperatures,

0°C is better defined as the melting point of ice. In this scale a temperature difference of 1 degree Celsius is the same as a 1kelvin increment, but the scale is offset by the temperature at which ice melts (273.15 K).

By international agreement the Kelvin and Celsius scales are defined by two fixing points: absolute zero and the triple point of Vienna Standard Mean Ocean Water, which is water specially prepared with a specified blend of hydrogen and oxygen isotopes. Absolute zero is defined as precisely 0K and −273.15°C. It is the temperature at which all classical translational motion of the particles comprising matter ceases and they are at complete rest in the classical model. Quantum-mechanically, however, zero-point motion remains and has an associated energy, the zero-point energy. Matter is in its ground state, and contains no thermal energy. The triple point of water is defined as 273.16K and 0.01°C. This definition serves the following purposes: it fixes the magnitude of the kelvin as being precisely 1 part in 273.16 parts of the difference between absolute zero and the triple point of water; it establishes that one kelvin has precisely the same magnitude as one degree on the Celsius scale; and it establishes the difference between the null points of these scales as being 273.15K (0K = −273.15°C and 273.16K = 0.01°C).

In the United States, the Fahrenheit scale is widely used. On this scale the freezing point of water corresponds to 32 °F and the boiling point to 212 °F. The Rankine scale, still used in fields of chemical engineering in the U.S., is an absolute scale based on the Fahrenheit increment.

Conversion

The following table shows the temperature conversion formulas for conversions to and from the Celsius scale.

	from Celsius	**to Celsius**
Fahrenheit	$[°F] = [°C] \times \frac{9}{5} + 32$	$[°C] = ([°F] - 32) \times \frac{5}{9}$
Kelvin	$[K] = [°C] + 273.15$	$[°C] = [K] - 273.15$
Rankine	$[°R] = ([°C] + 273.15) \times \frac{9}{5}$	$[°C] = ([°R] - 491.67) \times \frac{5}{9}$
Delisle	$[°De] = (100 - [°C]) \times \frac{3}{2}$	$[°C] = 100 - [°De] \times \frac{2}{3}$
Newton	$[°N] = [°C] \times \frac{33}{100}$	$[°C] = [°N] \times \frac{100}{33}$
Réaumur	$[°Ré] = [°C] \times \frac{4}{5}$	$[°C] = [°Ré] \times \frac{5}{4}$
Rømer	$[°Rø] = [°C] \times \frac{21}{40} + 7.5$	$[°C] = ([°Rø] - 7.5) \times \frac{40}{21}$

Plasma Physics

The field of plasma physics deals with phenomena of electromagnetic nature that involve very high temperatures. It is customary to express temperature as energy in units of electronvolts (eV) or kiloelectronvolts (keV). The energy, which has a different dimension from temperature, is then calculated as the product of the Boltzmann constant and temperature, $E = k_B T$. Then, 1 eV corresponds to 11605K. In the study of QCD matter one routinely encounters temperatures of the order of a few hundred MeV, equivalent to about 10^{12}K.

Theoretical Foundation

Historically, there are several scientific approaches to the explanation of temperature: the classical thermodynamic description based on macroscopic empirical variables that can be measured in a laboratory; the kinetic theory of gases which relates the macroscopic description to the probability distribution of the energy of motion of gas particles; and a microscopic explanation based on statistical physics and quantum mechanics. In addition, rigorous and purely mathematical treatments have provided an axiomatic approach to classical thermodynamics and temperature. Statistical physics provides a deeper understanding by describing the atomic behavior of matter, and derives macroscopic properties from statistical averages of microscopic states, including both classical and quantum states. In the fundamental physical description, using natural units, temperature may be measured directly in units of energy. However, in the practical systems of measurement for science, technology, and commerce, such as the modern metric system of units, the macroscopic and the microscopic descriptions are interrelated by the Boltzmann constant, a proportionality factor that scales temperature to the microscopic mean kinetic energy.

The microscopic description in statistical mechanics is based on a model that analyzes a system into its fundamental particles of matter or into a set of classical or quantum-mechanical oscillators and considers the system as a statistical ensemble of microstates. As a collection of classical material particles, temperature is a measure of the mean energy of motion, called kinetic energy, of the particles, whether in solids, liquids, gases, or plasmas. The kinetic energy, a concept of classical mechanics, is half the mass of a particle times its speed squared. In this mechanical interpretation of thermal motion, the kinetic energies of material particles may reside in the velocity of the particles of their translational or vibrational motion or in the inertia of their rotational modes. In monatomic perfect gases and, approximately, in most gases, temperature is a measure of the mean particle kinetic energy. It also determines the probability distribution function of the energy. In condensed matter, and particularly in solids, this purely mechanical description is often less useful and the oscillator model provides a better description to account for quantum mechanical phenomena. Temperature determines the statistical occupation of the microstates of the ensemble. The microscopic definition of temperature is only meaningful in the thermodynamic limit, meaning for large ensembles of states or particles, to fulfill the requirements of the statistical model.

In the context of thermodynamics, the kinetic energy is also referred to as thermal energy. The thermal energy may be partitioned into independent components attributed to the degrees of freedom of the particles or to the modes of oscillators in a thermodynamic system. In general, the number of these degrees of freedom that are available for the equipartitioning of energy depend on the temperature, i.e. the energy region of the interactions under consideration. For solids, the thermal energy is associated primarily with the vibrations of its atoms or molecules about their equilibrium position. In an ideal monatomic gas, the kinetic energy is found exclusively in the purely translational motions of the particles. In other systems, vibrational and rotational motions also contribute degrees of freedom.

Kinetic Theory of Gases

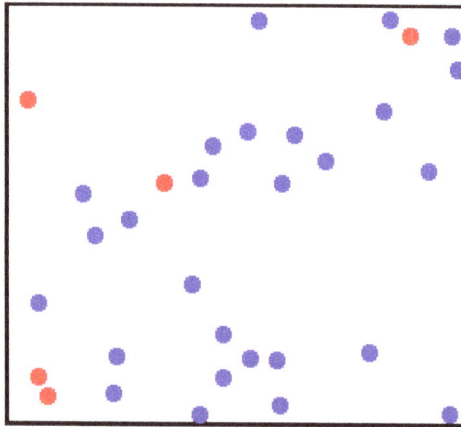

A theoretical understanding of temperature in an ideal gas can be obtained from the Kinetic theory.

Maxwell and Boltzmann developed a kinetic theory that yields a fundamental understanding of temperature in gases. This theory also explains the ideal gas law and the observed heat capacity of monatomic (or 'noble') gases.

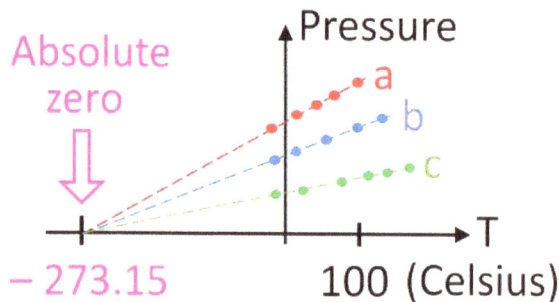

Plots of pressure vs temperature for three different gas samples extrapolated to absolute zero.

The ideal gas law is based on observed empirical relationships between pressure (p), volume (V), and temperature (T), and was recognized long before the kinetic theory of gases was developed. The ideal gas law states:

$$pV = nRT$$

where n is the number of moles of gas and $R = 8.3144598(48)$ J mol^{-1} K^{-1} is the gas constant.

This relationship gives us our first hint that there is an absolute zero on the temperature scale, because it only holds if the temperature is measured on an absolute scale such as Kelvins. The ideal gas law allows one to measure temperature on this absolute scale using the gas thermometer. The temperature in kelvins can be defined as the pressure in pascals of one mole of gas in a container of one cubic meter, divided by the gas constant.

Although it is not a particularly convenient device, the gas thermometer provides an essential theoretical basis by which all thermometers can be calibrated. As a practical matter it is not possible to use a gas thermometer to measure absolute zero temperature since the gases tend to condense into a liquid long before the temperature reaches zero. It is possible, however, to extrapolate to absolute zero by using the ideal gas law, as shown in the figure.

The kinetic theory assumes that pressure is caused by the force associated with individual atoms striking the walls, and that all energy is translational kinetic energy. Using a sophisticated symmetry argument, Boltzmann deduced what is now called the Maxwell–Boltzmann probability distribution function for the velocity of particles in an ideal gas. From that probability distribution function, the average kinetic energy, E_k (per particle), of a monatomic ideal gas is:

$$E_k = \frac{1}{2} m v_{rms}^2 = \frac{3}{2} kT,$$

where the Boltzmann constant, k, is the ideal gas constant divided by the Avogadro number, and v_{rms} is the root-mean-square speed. Thus the ideal gas law states that internal energy is directly proportional to temperature. This direct proportionality between temperature and internal energy is a special case of the equipartition theorem, and holds only in the classical limit of an ideal gas. It does not hold for most substances, although it is true that temperature is a monotonic (non-decreasing) function of internal energy.

Zeroth Law of Thermodynamics

When two otherwise isolated bodies are connected together by a rigid physical path impermeable to matter, there is spontaneous transfer of energy as heat from the hotter to the colder of them. Eventually they reach a state of mutual thermal equilibrium, in which heat transfer has ceased, and the bodies' respective state variables have settled to become unchanging.

One statement of the zeroth law of thermodynamics is that if two systems are each in thermal equilibrium with a third system, then they are also in thermal equilibrium with each other.

This statement helps to define temperature but it does not, by itself, complete the definition. An empirical temperature is a numerical scale for the hotness of a thermodynamic system. Such hotness may be defined as existing on a one-dimensional manifold, stretching between hot and cold. Sometimes the zeroth law is stated to include the existence of a unique universal hotness manifold, and of numerical scales on it, so as to provide a complete definition of empirical temperature. To be suitable for empirical thermometry, a material must have a monotonic relation between hotness and some easily measured state variable, such as pressure or volume, when all other relevant coordinates are fixed. An exceptionally suitable system is the ideal gas, which can provide a temperature scale that matches the absolute Kelvin scale. The Kelvin scale is defined on the basis of the second law of thermodynamics.

Negative Temperature

On the empirical temperature scales, which are not referenced to absolute zero, a negative temperature is one below the zero-point of the scale used. For example, dry ice has a sublimation temperature of $-78.5°C$ which is equivalent to $-109.3°F$. On the absolute Kelvin scale, however, this temperature is 194.6 K. On the absolute scale of thermodynamic temperature no material can have a temperature smaller than or equal to 0 K, both of which are forbidden by the third law of thermodynamics.

Temperature is basically defined for a body in its own state of internal thermodynamic equilibrium, and in this definition, on an absolute scale, it is always positive. In an apparent contradiction of this reliable and valid rule, a so-called negative absolute "temperature" may be approximately defined for a component of a body that is not in its own state of internal thermodynamic equilibrium: a component may have a negative approximate "temperature" while the rest of the components of the body have positive approximate temperatures. Such a non-equilibrium situation is either transient in time or is maintained by external factors that drive a flow of energy through the body of interest. An example of such a component is a spin system within a body, as follows.

In the quantum mechanical description of electron and nuclear spin systems that have a limited number of possible states, and therefore a discrete upper limit of energy they can attain, it is possible to obtain a negative temperature, which is numerically indeed less than absolute zero. However, this is not the macroscopic temperature of the material, but instead the temperature of only very specific degrees of freedom, that are isolated from others and do not exchange energy by virtue of the equipartition theorem.

A negative temperature is experimentally achieved with suitable radio frequency techniques that cause a population inversion of spin states from the ground state. As the energy in the system increases upon population of the upper states, the entropy increases as well, as the system becomes less ordered, but attains a maximum value when the spins are evenly distributed among ground and excited states, after which it begins to decrease, once again achieving a state of higher order as the upper states begin to fill exclusively. At the point of maximum entropy, the temperature function shows the behavior of a singularity, because the slope of the entropy function decreases to zero at first and then turns negative. Since temperature is the inverse of the derivative of the entropy, the temperature formally goes to infinity at this point, and switches to negative infinity as the slope turns negative. At energies higher than this point, the spin degree of freedom therefore exhibits formally a negative thermodynamic temperature. As the energy increases further by continued population of the excited state, the negative temperature approaches zero asymptotically. As the energy of the system increases in the population inversion, a system with a negative temperature is not colder than absolute zero, but rather it has a higher energy than at positive temperature, and may be said to be in fact hotter at negative temperatures. When brought into contact with a system at a positive temperature, energy will be transferred from the negative temperature regime to the positive temperature region.

Examples of Temperature

	Temperature		Peak emittance wavelength of black-body radiation
	Kelvin	**Celsius**	
Absolute zero (precisely by definition)	0 K	−273.15 °C	cannot be defined
Coldest temperature achieved	100 pK	−273.149999999900 °C	29,000 km
Coldest Bose–Einstein condensate	450 pK	−273.14999999955 °C	6,400 km
One millikelvin (precisely by definition)	0.001 K	−273.149 °C	2.89777 m (radio, FM band)
Cosmic microwave background (2013 measurement)	2.7260 K	−270.424 °C	0.00106301 m (millimeter-wavelength microwave)
Water triple point (precisely by definition)	273.16 K	0.01 °C	10,608.3 nm (long-wavelength IR)
Water boiling point[A]	373.1339 K	99.9839 °C	7,766.03 nm (mid-wavelength IR)
Incandescent lamp[B]	2500 K	≈2,200 °C	1,160 nm (near infrared)[C]
Sun's visible surface[D]	5,778 K	5,505 °C	501.5 nm (green-blue light)
Lightning bolt channel[E]	28 kK	28,000 °C	100 nm (far ultraviolet light)
Sun's core[E]	16 MK	16 million °C	0.18 nm (X-rays)

Thermonuclear weapon (peak temperature)[E]	350 MK	350 million °C	8.3×10^{-3} nm (gamma rays)
Sandia National Labs' Z machine[E]	2 GK	2 billion °C	1.4×10^{-3} nm (gamma rays)[F]
Core of a high-mass star on its last day[E]	3 GK	3 billion °C	1×10^{-3} nm (gamma rays)
Merging binary neutron star system[E]	350 GK	350 billion °C	8×10^{-6} nm (gamma rays)
Relativistic Heavy Ion Collider[E]	1 TK	1 trillion °C	3×10^{-6} nm (gamma rays)
CERN's proton vs nucleus collisions[E]	10 TK	10 trillion °C	3×10^{-7} nm (gamma rays)
Universe 5.391×10^{-44} s after the Big Bang[E]	1.417×10^{32} K	1.417×10^{32} °C	1.616×10^{-27} nm (Planck length)

- A For Vienna Standard Mean Ocean Water at one standard atmosphere (101.325 kPa) when calibrated strictly per the two-point definition of thermodynamic temperature.

- B The 2500 K value is approximate. The 273.15 K difference between K and °C is rounded to 300 K to avoid false precision in the Celsius value.

- C For a true black-body (which tungsten filaments are not). Tungsten filament emissivity is greater at shorter wavelengths, which makes them appear whiter.

- D Effective photosphere temperature. The 273.15 K difference between K and °C is rounded to 273 K to avoid false precision in the Celsius value.

- E The 273.15 K difference between K and °C is within the precision of these values.

- F For a true black-body (which the plasma was not). The Z machine's dominant emission originated from 40 MK electrons (soft x-ray emissions) within the plasma.

References

- Bird, R. Byron; Stewart, Warren E.; Lightfoot, Edwin N. (2007-01-01). Transport Phenomena. John Wiley & Sons. ISBN 9780470115398.

- Sam Zhang; Dongliang Zhao (19 November 2012). Aeronautical and Aerospace Materials Handbook. CRC Press. pp. 304–. ISBN 978-1-4398-7329-8. Retrieved 7 May 2013.

- George E. Totten (2002). Handbook of Residual Stress and Deformation of Steel. ASM International. pp. 322–. ISBN 978-1-61503-227-3. Retrieved 7 May 2013.

- Rajiv Asthana; Ashok Kumar; Narendra B. Dahotre (9 January 2006). Materials Processing and Manufacturing Science. Butterworth–Heinemann. pp. 158–. ISBN 978-0-08-046488-6. Retrieved

- John J. Lentini - Scientific Protocols for Fire Investigation, CRC 2006, ISBN 0849320828, table from NFPA 921, Guide for Fire and Explosion Investigations

- Paul A., Tipler; Gene Mosca (2008). Physics for Scientists and Engineers, Volume 1 (6th ed.). New York, NY: Worth Publishers. pp. 666–670. ISBN 1-4292-0132-0.

- Varshneya, A. K. (2006). Fundamentals of inorganic glasses. Sheffield: Society of Glass Technology. ISBN 0-12-714970-8.

- Raymond Serway; John Jewett (2005), Principles of Physics: A Calculus-Based Text, Cengage Learning, p. 506, ISBN 0-534-49143-X

- Kittel, Charles (2005). Introduction to Solid State Physics (8th ed.). Hoboken, New Jersey, USA: John Wiley & Sons. p. 141. ISBN 0-471-41526-X.

- Blundell, Stephen (2001). Magnetism in Condensed Matter. Oxford Master Series in Condensed Matter Physics (1st ed.). Hoboken, New Jersey, USA: Oxford University Press. p. 27. ISBN 978-0-19-850591-4.

- Kittel, Charles (2005). Introduction to Solid State Physics (8th ed.). Hoboken, New Jersey, USA: John Wiley & Sons. p. 141. ISBN 0-471-41526-X.

- Münster, A. (1970), Classical Thermodynamics, translated by E.S. Halberstadt, Wiley–Interscience, London, ISBN 0-471-62430-6, pp. 49, 69.

- Bailyn, M. (1994). A Survey of Thermodynamics, American Institute of Physics Press, New York, ISBN 0-88318-797-3, pp. 14–15, 214.

- Callen, H.B. (1960/1985), Thermodynamics and an Introduction to Thermostatistics, (first edition 1960), second edition 1985, John Wiley & Sons, New York, ISBN 0-471-86256-8, pp. 146–148.

- Kondepudi, D., Prigogine, I. (1998). Modern Thermodynamics. From Heat Engines to Dissipative Structures, John Wiley, Chichester, ISBN 0-471-97394-7, pp. 115–116.

- Glansdorff, P., Prigogine, I., (1971). Thermodynamic Theory of Structure, Stability and Fluctuations, Wiley, London, ISBN 0-471-30280-5, pp. 14–16.

- Bailyn, M. (1994). A Survey of Thermodynamics, American Institute of Physics Press, New York, ISBN 0-88318-797-3, pp. 133–135.

- Callen, H.B. (1960/1985), Thermodynamics and an Introduction to Thermostatistics, (first edition 1960), second edition 1985, John Wiley & Sons, New York, ISBN 0-471-86256-8, pp. 309–310.

- Serrin, J. (1986). Chapter 1, 'An Outline of Thermodynamical Structure', pages 3-32, especially page 6, in New Perspectives in Thermodynamics, edited by J. Serrin, Springer, Berlin, ISBN 3-540-15931-2.

- M. J. Moran, H. N. Shapiro (2006). "1.6.1". Fundamentals of Engineering Thermodynamics (5 ed.). John Wiley & Sons, Ltd. p. 14. ISBN 978-0-470-03037-0.

- Kittel, Charles; Kroemer, Herbert (1980). Thermal Physics (2nd ed.). W. H. Freeman Company. pp. 391–397. ISBN 0-7167-1088-9.

Permissions

All chapters in this book are published with permission under the Creative Commons Attribution Share Alike License or equivalent. Every chapter published in this book has been scrutinized by our experts. Their significance has been extensively debated. The topics covered herein carry significant information for a comprehensive understanding. They may even be implemented as practical applications or may be referred to as a beginning point for further studies.

We would like to thank the editorial team for lending their expertise to make the book truly unique. They have played a crucial role in the development of this book. Without their invaluable contributions this book wouldn't have been possible. They have made vital efforts to compile up to date information on the varied aspects of this subject to make this book a valuable addition to the collection of many professionals and students.

This book was conceptualized with the vision of imparting up-to-date and integrated information in this field. To ensure the same, a matchless editorial board was set up. Every individual on the board went through rigorous rounds of assessment to prove their worth. After which they invested a large part of their time researching and compiling the most relevant data for our readers.

The editorial board has been involved in producing this book since its inception. They have spent rigorous hours researching and exploring the diverse topics which have resulted in the successful publishing of this book. They have passed on their knowledge of decades through this book. To expedite this challenging task, the publisher supported the team at every step. A small team of assistant editors was also appointed to further simplify the editing procedure and attain best results for the readers.

Apart from the editorial board, the designing team has also invested a significant amount of their time in understanding the subject and creating the most relevant covers. They scrutinized every image to scout for the most suitable representation of the subject and create an appropriate cover for the book.

The publishing team has been an ardent support to the editorial, designing and production team. Their endless efforts to recruit the best for this project, has resulted in the accomplishment of this book. They are a veteran in the field of academics and their pool of knowledge is as vast as their experience in printing. Their expertise and guidance has proved useful at every step. Their uncompromising quality standards have made this book an exceptional effort. Their encouragement from time to time has been an inspiration for everyone.

The publisher and the editorial board hope that this book will prove to be a valuable piece of knowledge for students, practitioners and scholars across the globe.

Index

www.ingramcontent.com/pod-product-compliance
Lightning Source LLC
Chambersburg PA
CBHW061932190326
41458CB00009B/2724